Library of Congress Cataloging-in-Publication Data
Ede, Andrew.
   The chemical element : a historical perspective/Andrew Ede.
       p.   cm. — (Greenwood guides to great ideas in science, ISSN 1559–5374))
   Includes bibliographical references and index.
   ISBN 0–313–33304–1 (alk. paper)
   1. Chemical elements—History.   2. Matter—Constitution—History.
I. Title. II. Series.
   QD466.E34   2006
   546—dc22   2006004987

British Library Cataloguing in Publication Data is available.

Library of Congress Catalog Card Number: 2006004987
ISBN: 0–313–33304–1
ISSN: 1559–5374
First published in 2006

Greenwood Press, 88 Post Road West, Westport, CT 06881
An imprint of Greenwood Publishing Group, Inc.
www.greenwood.com

Printed in the United States of America

CONTENTS

Series Foreword   *vii*

List of Illustrations   *ix*

Introduction   *xi*

1	Ancient Matter Theory	1
2	Greek Matter Theory from Thales to Aristotle	9
3	Matter Theory in China, the Middle East, and India	19
4	Alchemy and the Concept of Elements	31
5	Corpuscles and Atoms	45
6	Bringing Order to Chaos	57
7	Making Elements Elemental	67
8	Seeking Order: The Periodic Table	73
9	The Atomic Elements	87
10	The Elements beyond 92	105

Conclusion   *111*

Appendix 1: The Periodic Table   *113*

Appendix 2: History of the Elements   *115*

Appendix 3: The Elements by Alphabetical Order   *157*

Appendix 4: The Elements by Date of Discovery   *159*

Glossary   *161*

Annotated Bibliography   *167*

Bibliography   *171*

Index   *175*

SERIES FOREWORD

The volumes in this series are devoted to concepts that are fundamental to different branches of the natural sciences—the gene, the quantum, geological cycles, planetary motion, evolution, the cosmos, and forces in nature, to name just a few. Although these volumes focus on the historical development of scientific ideas, the underlying hope of this series is that the reader will gain a deeper understanding of the process and spirit of scientific practice. In particular, in an age in which students and the public have been caught up in debates about controversial scientific ideas, it is hoped that readers of these volumes will better appreciate the provisional character of scientific truths by discovering the manner in which these truths were established.

The history of science as a distinctive field of inquiry can be traced to the early seventeenth century when scientists began to compose histories of their own fields. As early as 1601, the astronomer and mathematician, Johannes Kepler, composed a rich account of the use of hypotheses in astronomy. During the ensuing three centuries, these histories were increasingly integrated into elementary textbooks, the chief purpose of which was to pinpoint the dates of discoveries as a way of stamping out all too frequent propriety disputes, and to highlight the errors of predecessors and contemporaries. Indeed, historical introductions in scientific textbooks continued to be common well into the twentieth century. Scientists also increasingly wrote histories of their disciplines—separate from those that appeared in textbooks—to explain to a broad popular audience the basic concepts of their science.

The history of science remained under the auspices of scientists until the establishment of the field as a distinct professional activity in middle of the twentieth century. As academic historians assumed control of history of science writing, they expended enormous energies in the attempt to forge a distinct and autonomous discipline. The result of this struggle to position the history of science as an intellectual endeavor that was valuable in its own right, and not merely in consequence of its ties to science, was that historical studies of the natural sciences were no longer composed with an eye toward educating a wide audience that included non-scientists, but instead were composed with the aim of being consumed by other professional historians of science. And as historical breadth was sacrificed for technical detail, the literature became increasingly daunting in its technical detail. While this scholarly work increased our understanding of the nature of science, the technical demands imposed on the reader had the unfortunate consequence of leaving behind the general reader.

As Series Editor, my ambition for these volumes is that they will combine the best of these two types of writing about the history of science. In step with the general introductions that we associate with historical writing by scientists, the purpose of these volumes is educational; they have been authored with the aim of making these concepts accessible to students—high school, college, and university—and to the general public. However, the scholars who have written these volumes are not only able to impart genuine enthusiasm for the science discussed in the volumes of this series, they can use the research and analytic skills that are the staples of any professional historian and philosopher of science to trace the development of these fundamental concepts. My hope is that a reader of these volumes will share some of the excitement of these scholars—for both science, and its history.

Brian Baigrie
University of Toronto
Series Editor

LIST OF ILLUSTRATIONS

1.	Egyptian god Osiris.	3
2.	Geometric solids.	11
3.	Zeno's paradox.	12
4.	Aristotle's model of the Universe.	15
5.	Aristotle's Four Elements and Four Qualities.	15
6.	Yin Yang symbol.	21
7.	Alchemical symbols for materials and planets.	36
8.	The Alchemist by David Teniers the Younger, c. 1645.	40
9.	Robert Boyle's air pump.	53
10.	Joseph Priestley's pneumatic trough used for collecting gases.	55
11.	Lavoisier's constant pressure gas pump and reaction vessels.	59
12.	The calorimeter.	62
13.	Lavoisier's table of simple substances from *Elements of Chemistry* (1790).	63
14.	Dalton's elements and common "atoms." From John Dalton, *New System of Chemical Philosophy* (1808).	69
15.	Gustav Kirchhoff and Robert Bunsen's spectrocope, from "Chemical analysis by Observation of Spectra," (1860).	76
16.	Mendeleev's periodic table from *Annalen der Chemie,* (1872).	81
17.	Ceria and yttria isolation.	89
18.	Ultramicroscope.	94
19.	Fission.	101
20.	Fusion.	107

INTRODUCTION

The study of matter is fundamental to human survival, and it is linked forever to civilization. On a practical level, humans have come to dominate nature because we can identify, manipulate, and transform the materials of the world. We take a stone and make an axe, then use the axe to chop down trees, transforming them into firewood and lumber to make houses. Given the importance of the material world, it is not surprising that humans have spent a great deal of time trying to determine what matter is and how it works.

The quest to understand the physical material of the world has gone through many stages, but they are all part of a broader idea called matter theory. Matter theory covers the changing ideas and systems that were used to describe and explain the material world. A large part of matter theory was based on a theory of the elements. The precise definition of "element" has changed radically over time, but it has always been referred to the most basic type of material. The effort to find the elements is one of the most dramatic stories in the development of civilization, and it was in large part the interest in the elements that led to the creation of modern chemistry.

The first step toward understanding matter was classification. We identified things like rocks, trees and plants, animals and water, and then grouped them in various ways, classifying them as solid or liquid, alive or not alive, edible or inedible. Although the naming and grouping of things that people saw around them was important (and continues to this day), classification was not enough, because almost as soon as people began manipulating matter, they discovered that it could be transformed. This led to a profound and complex philosophical question: If one thing can be transformed into another thing, what then is its true nature? For example, water, ice, and steam all seem to be related, but are they three substances or one substance with three radically different characteristics?

Although humans have been manipulating matter for tens of thousands of years, a great technical leap forward occurred around 5000 B.C.E., when the

Egyptians began smelting copper. The discovery of smelting was probably accidental, but once the principle was understood, it opened the door to a whole new range of tools and objects, from cups to jewelry. To make copper, you take a particular bluish-green rock and heat it in a fire. After a short time, you get a bright liquid that cools into a shiny, malleable metal. This miraculous transformation produces something that can be stretched, molded, hammered, and bent into an infinite number of shapes. And, as amazing as it is, the new material has one more property that makes it special. It can be heated and melted over and over without changing its properties. Given smelting's ability to effect this transformation, it is easy to understand why it was considered magical and the process was often a carefully guarded trade secret. It is also easy to see why smelting raised philosophical questions about the nature of matter, since the characteristics of the smelted copper seemed permanent, while those of the original copper ore clearly were not. This finding suggested that copper was more basic than the ore and that something about the process of smelting transformed the ore into metal.

Smelting was not the only transformation that ancient people tried to understand. Growth and decay, combustion, oxidization (or rusting), and fermentation were all significant transformations of one material into another. One of the most important transformations was the production of salt from brine, which produced a product so necessary to civilization that few of us would be here today if salt production had not been figured out. A large part of the history of science comes directly from efforts to make sense of these transformations and to determine what was the basic or purest elemental material.

Our modern definition of an element is a combination of physical and chemical ideas that have been worked out over thousands of years. We now define a chemical element as a substance that cannot be decomposed or broken down into simpler substances by chemical means. This is a logical and useful definition, but it took generations to establish experimentally. By long experience running back to the dawn of civilization, certain substances, such as copper, gold, and water, came to be considered pure, but whether these were actually elements or just materials that were very hard to decompose was not entirely clear. Early natural philosophers argued that there had to be a very small number of true elements, perhaps even just a single prime element, and that all the matter of the world was made by combining the true element or elements in particular patterns.

When the study of matter became more systematic, the number of known elements started to rise, from the handful of "pure" substances known to ancient people to the dozens recognized by the time of Antoine Lavoisier in the eighteenth century. Lavoisier and his followers reformed chemistry, partly on the basis of detailed work to clarify the definition of what an element was and partly through careful experiments to identify the characteristics of the known elements. While this work was vital, it actually complicated the situation, as the list of elements continued to grow. Most chemists felt that there had to be some system behind the existence of so many elements, but no one could figure out how the elements were related until Dimitri Mendeleev and Lothar Meyer, working independently, produced periodic tables of the elements, demonstrating the underlying relationships of the basic materials of the world.

The periodic table of elements was a great development in science, acting as a system of classification, a tool for research, and a guide to further study. The earliest tables seemed to suggest that there was order in the elements, but they also contained errors that limited their utility. This changed when Dimitri Mendeleev created a table that had gaps representing elements that had not been found. Extrapolating from the characteristics for chemical groups as demonstrated by the table of elements, chemists had sense of what to look for and worked to fill in the blank spots. Even more important was the direction the tables gave to a closer examination of the atom. The elements were grouped by mass (a measurable physical characteristic) and by chemical behavior. For example, some elements, such as oxygen and hydrogen, combined with many other elements, while neon and gold did not seem to combine with anything else, and their groupings reflected these distinctions. To understand this range of behavior, scientists needed to look more closely at the basic unit of the elements—the atom. Then an unexpected discovery changed everyone's ideas about the smallest particles. In 1897, the electron was discovered, and atoms turned out to be made up of even smaller particles, now known as neutrons, protons, and electrons.

The new work led to a physical definition of an element based on atomic structure. According to the physical definition of an element, all the atoms of a particular element have the same number of protons and electrons, although the number of neutrons may vary. The placement of the protons, electrons, and neutrons in the atom follow strict rules of arrangement. The number of subatomic particles and their arrangement then determines the way atoms of elements interact.

Electrons are arranged in shells around the nucleus. Each shell contains a certain number of electrons, and there is a maximum number for each shell.

The shapes of the orbits and the determination of how many electrons actually can occupy a shell are very complex. At the most basic level, the bonding of atoms into molecules is based on a tendency of any atom to try to fill a shell to the maximum number of electrons by sharing electrons. Thus, a carbon atom, which has two electrons in the inner shell (the maximum possible) and four electrons in the outer (looking for four more), will bond with four hydrogen atoms, each of which has one in its only shell and is thus looking for one more to reach its maximum.

While the question of what elements were seemed to be solved with modern chemistry and physics, the question of how the elements came to be was still

Shell	1	2	3	4	5
Electrons (maximum)	2	8	18	32	50

unsolved until scientists began to explore the forces that make atoms possible. The quest to understand those forces took scientists into the heart of the atom, but it also linked matter to the biggest objects in the universe. The forces necessary to make the elements exist naturally in only one place: the nuclear furnace of stars. Every element, with the exception of elemental hydrogen, was created by the stars. The air we breathe, the grass and the trees, and our own bodies are composed of matter created by the life and death of stars.

Scientists continue to work on the problem of the elements, refining what we know about existing elements and searching for new ones. Several super-heavy elements have been created in the laboratory. While these elements can exist only for fractions of a second, they are telling us more about the way the universe operates. New materials such as nano carbon tubes and solid-state circuits exist because of our greater understanding of matter that came from research on the elements.

The history of the elements, and the history of the concept of the elements, parallels the history of civilization. From the ancient stories of creation to the cutting edge of modern high-energy physics, the study of the material world has shaped the course of empires, given us power over nature, and been one of the keys to the development of science. Through the ages, many of the great thinkers, such as Aristotle, René Descartes, and Isaac Newton, grounded their understanding of the universe in matter theory. Understanding the struggles and stories of those who searched for the elements brings us closer to our heritage and offers us insight into why the world is the way it is today.

A BRIEF NOTE ON DATES AND NAMES

The text follows the Western system of dating, with dates given in the form B.C.E. for "Before Current Era" and C.E. for "Current Era," rather than B.C. and A.D. The era indicator is used for all ancient dates and early current-era dates prior to 1000 C.E. Dates that appear with no era indicator are assumed to be current era. In cases were the exact date in unknown or unclear, a "c." is used, meaning "circa" or "fl.," meaning "flourished."

Names are given in their most common English form. Since many names are translated from other languages, there may be alternative spellings. For people who contributed to the development of the concept of the elements, the text includes the birth and death dates after the first mention of the person.

1

ANCIENT MATTER THEORY

The ancient origins of matter theory are really the origins of human ideas about the making of the world. For as long as humans have tried to explain why the world exists and how the parts of it are put together, we have been creating systems to identify, classify, and explain matter. Almost all creation stories include a conception of the elements, or those types of matter that were the original material from which all of the rest of the world was constructed. A look at these early theories reveals our ongoing interest in matter. Linking matter to creation also tells us about the sense that many people had that the study of matter was also a study of the divine. The idea that understanding matter would lead to an understanding of something secret, supernatural, or divine has been a part of matter theory until almost the present day.

Human ideas about the matter that makes up the world have developed in conjunction with our ability to transform and manipulate matter. We discovered the elements by trying to understand how we could do more with the material around us. These efforts go back to the very origin of human life. The manipulation of matter has been so important to human life that historians and archeologists have often identified different periods in history by the most advanced material a particular society could produce at a given time. Thus, we have the Paleolithic era, or Stone Age, followed by the Bronze Age and the Iron Age. The oldest stone tools, found in 1997 near the Gona River, in Ethiopia, by the researchers John W. K. Harris and Sileshi Semaw, date from 2.5 to 2.6 million years ago. Stone tools are the primary record of our hominid ancestors' attempts to modify the environment. Tool use almost certainly predates this time, but objects made of other materials, such as wood, bone, or horn, have not survived the passing of the ages or cannot be clearly shown to have been tools.

When our ancestors began turning sticks, bones, and rocks (especially volcanic rocks like flint and cert) into tools, they began to reshape nature to suit their needs. They were also learning the characteristics of different types of matter. Over time, specific observations became general categories. For example, the concepts of liquid and solid were separated from particular objects such as water and rock. This contributed a vocabulary of abstract terms that could then be used to talk about the nature of things, rather than just the things themselves.

One of the most profound aspects of ancient matter theory was its connection to creation stories. Almost all ancient societies share a common bond in explaining the origins of the universe as a story about the creation of matter from nothingness. There are usually three (water, land, and air) or four elements (water, land, air, and fire) in the first instance of creation. For example, in the Hopi tradition, Taiowa, the Creator, existed in the infinite. The infinite had no shape, time, or life except the mind of the creator. Taiowa created Sotuknang and directed Sotuknang to make from the endless space nine solid worlds. Into these worlds were added water and air. Once these were created, life could be made.

In the Hindu creation story of Praja-pati, only thought existed in nothingness. The thought thought "Let me have a self," and a mind was created. As the mind (sometimes identified as Brahma) moved about in worship, water was created. The motion of the mind in the water made a froth, which solidified into earth. As he lay upon the earth, his luminescence became fire. After resting, he split into three parts: fire, the sun, and the air.

The story of Pan Gu, from China, even includes a physical explanation for the place of matter in the world. In the beginning, the universe was contained in an egg-shaped cloud. All matter was mixed and in chaos. At the center of the egg lay Pan Gu, a sleeping giant. When he awoke after thousands of years of sleep, he broke the egg apart, releasing the matter of the universe. The light, pure elements flew up and became the sky and the heavens, and the heavy, impure elements descended and became the earth. Fearing that the elements would remix and chaos return, Pan Gu held the earth and the sky apart. When the world was stable and did not need to be held apart, Pan Gu died, and his body was transformed into the mountains, rivers, and winds of the world.

In the Egyptian creation stories, the sun-god Atum produced from himself Shu, the god of air, and Tefnut, the goddess of moisture. Shu and Tefnut mated and produced the earth and the sky, as well as the earth-god Geb and the sky-goddess Nut. From the mating of Geb and Nut came the pantheon of Egyptian gods, including Osiris and Isis, Seth and Nephthys.

The creation stories contain far more than matter theory, but they share some characteristics that tell us about the human drive to understand matter. In most stories, matter does not exist until called into being by a creator. The first matter created is usually completely pure, untainted by anything else and made up of only one undifferentiated substance. Thus, the start of the physical universe begins when the creator introduces the first element. The prime element must then be divided or transformed into classes of matter, typically earth, water, air, and sometimes fire or light. There is almost always a sense that the first matter is more malleable than matter in the present age. The rest of the world is then created by mixing the original elements. In some cases, the other objects, especially plants, animals, and people, are created whole (although still a mixture of the pure elements), such as in the biblical story of Genesis, while in others they are grown from the original elements or transformed from something else, such as in the Pan Gu story.

Another philosophical idea common to the presentation of matter in creation stories is that matter is the opposite of chaos. Matter is thus intimately linked to the human concept of an underlying order to nature. Wherever we look, we seem to see matter demonstrating order, whether it is the regularity of a leaf on a tree, our own ten fingers and toes, or the shape of crystals. It is not surprising that the study of matter has also been the search for order, both in the direct sense of how the bits go together to make a tree or a person

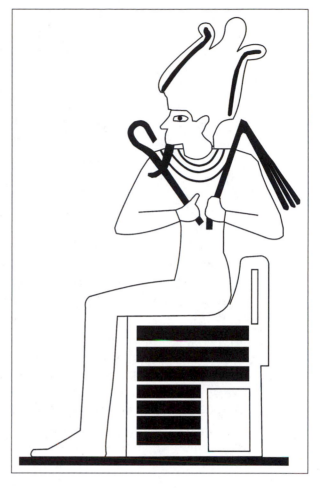

Figure 1: Egyptian god Osiris.

and also in the abstract sense of underlying patterns and principles that make it possible for a tree to exist at all.

While creation stories give us an insight into the philosophical foundation of our interest in matter and the elements, there were also very practical reasons to investigate matter. Our ability to manipulate physical objects has allowed us to prosper in a world where our innate physical abilities are very limited. Lacking fangs, claws, tough hides, or acute senses, humans are generally ill equipped to deal with the rigors of nature. Our strategy has been to work cooperatively and to make tools to overcome our lack of physical abilities. One of the greatest skills we learned early in human history was the controlled use of fire. The earliest evidence of the controlled use of fire comes from the discovery, by C. K. Brain and A. Sillent, of burnt bone fragments in the Swartkrans cave in South Africa. Tests on the bones indicated that they were between 1 and 1.5 million years old. Since this site was used by Homo erectus long before the appearance of Homo sapiens, our ancestors may have inherited or stolen the secret of fire from our relatives. It now seems likely that controlled use of fire has always been a part of human life.

Fire was the great transformer. It transformed materials, and it transformed our relationship with the material world. The first uses of fire were probably the obvious ones—for warmth and light. The use of fire for cooking and for processing materials followed and was likely the result of a combination of accident and experimentation driven by human curiosity to see what would happen when things were put in the fire. The introduction of cooking provided a number of advantages, such as making food, especially meat, easier to eat and less dangerous by destroying pathogens (although the scientific reason for this benefit was not understood for several thousand years). Tough foods like grains and woody root vegetables were softened by boiling, creating an ancient version of porridge. Fire was also used to preserve food through drying and smoking. Putting parts of animals into the fire led to the discovery of the combustibility of animal fats, and the use of stone lamps fueled with fat appeared around 79,000 years ago.

Although sun-dried clay vessels had probably existed for some time, the introduction of fired ceramic pots, created by baking clay in hot ashes, appeared around 20,000 B.C.E. The fired pot was stronger and less porous than pots made of dried clay. Pots could be used for storage, cooking, and even holding fire. So useful were ceramic pots and jars that they were, along with clay bricks, the first mass-produced objects.

Given all the things that fire gave humans, it is not surprising that fire was an important part of early matter theory. Even if fire was part of a creation myth as one of the original elements, it was often also introduced to humans as a special gift or tool. The god Prometheus, according to the Greek legend, stole fire from the ruling god, Zeus, and gave it to the mortals. With fire at their disposal, the mortals created civilization. For his hubris, Prometheus was chained to a rock and tormented by an eagle that tore his flesh and ate his liver each day. Each night, his wounds healed, and for generations the torture continued, until Herakles (better known to us as Hercules) freed him. Prometheus also gave humans other gifts, such as writing, astronomy, and mining, but his greatest gift was fire. The story is deeply symbolic, containing warnings about the dangers of challenging the gods and the risk of using powerful items like fire.

Although the Prometheus story linked fire to civilization, fire was even more closely associated with life, since most things that were alive were warm and required warmth. If plants or animals got too cold, they died. Fire seemed to be alive, in constant motion and requiring food in the form of fuel to continue to exist. If the fuel ran out, the fire "died," just as a person without food dies. For these reasons, fire has always been regarded differently from the other elements. It has often been placed in a separate category of matter from earth, water, and air.

The dual aspects of destruction and creation associated with fire appear in the Japanese story of creation. In the beginning time, the birth of Kagutsuchi-no-Kami, the spirit or deity of fire, destroyed his mother, Izanami, but in the prolonged process of dying, she produced Kanayama-biko and Kanayama-hime, the god and goddess of metals, and Haniyasu-hiko and Haniyasu-hime, the god and goddess of earth. Thus, the death of Izanami was the beginning of mining, smelting, and metalwork, all intimately related to the sacred tradition of sword making.

Thus, in Japanese mythology, fire also introduced humanity to metal. This reflected the probable course of actual history. Although it seems likely that small amounts of pure copper, silver, and gold were collected from surface outcroppings, the mining of metallic ores did not start in earnest until it was possible to refine ore through melting and separation of metals from the ore-bearing rock. The easiest metals to smelt were copper, silver, and gold, and, because of their properties, these metals were at the heart of our interest in the elements. Of the three, copper was the most significant in terms of use by early civilizations. The collecting of surface rocks that contain high levels of metallic copper probably goes back thousands of years into prehistoric times; we have evidence that copper collection was practiced at least as far back as 7000 B.C.E. Large-scale underground mining of copper in the Middle East and in Afghanistan can be dated to at least 1400 to 1200 B.C.E., when mines in the Timna Valley were being excavated during the time of the pharaohs.

At first, the attraction of copper, silver, and gold was their luster. The shiny metals were attractive and were used in jewelry and decoration. Raw copper in its metallic state was scarce, but around 5000 B.C.E., people in the Middle East discovered that copper could be smelted from ore such as malachite. These blue-green rocks were far more abundant than metallic copper and produced copper when heated in a charcoal fire. The creation of smelting represents one of the greatest discoveries of all time. The earliest evidence of large-scale smelting comes from Khirbat Hamra Ifdan, in the Faynan district of modern Jordan. This site was a major production facility as early as 4600 B.C.E. Other sites in ancient Egypt and Sumeria produced hundreds of tons of copper starting around 4500 B.C.E. By 3000 B.C.E., lead and tin were also being smelted in Egypt and Babylon.

The Middle East was one of the great centers of mining, smelting, and metalwork, and thus also of matter theory. It is no surprise that these activities were important in the Judeo-Christian account of creation and the stories of the Bible. Genesis starts with the creation of the two realms, the heavens and the earth. By the end of the first day, there was light, water, and earth, and from these came the rest of the material world, including plants, animals, and people. The concept of transmutation also appears in the Old Testament, particularly in the dramatic transformation of Lot's wife into salt. This concept would have many implications for the creation of alchemy in a later period. The place of metalwork was also important, and in Numbers, chapter 31, most of the smelted metals are named when the priest Eleazar spoke to the men of war, listing gold, silver, brass, iron, tin, and lead.

The process of smelting seemed like a kind of magic. By heating bits of rock, people could produce a shiny, malleable material. Things like clay, wood, or stone could be shaped only once, their properties set and unchangeable, but copper or silver could be returned to its liquid state and then be made into something new with no loss of quality. Metals had some amazing properties. They could be bent, stretched, hammered, and shaped, as well as cast in molds. While those characteristics were very useful, the metallic elements had another property that made them special. They could also be infinitely recycled. They could be transformed by being heated over and over again. With its use in these magical qualities, it is not surprising that smelting was often a carefully guarded secret, but as the secret of mining and smelting spread, copper tools became more common.

Copper was an extremely useful material, but it was too soft for some applications, particularly weapons and cutting tools. Some time around 3500 B.C.E., bronze was discovered. The discovery was probably the result of trials made with copper to change its properties. The addition of a small amount of tin (from 1% to 30% of the total metal) to copper produced a more rigid and tougher material. Bronze became the most important metal product available, used in everything from statues to cups, toga pins to swords. Although there is debate about whether knowledge of bronze production was carried to other parts of the world or whether it was independently discovered in other centers of copper production, such as China and India, the discovery that a mixture of copper and tin produced a tough and useful new material was so significant that we call the period the Bronze Age after the alloy.

The creation of alloys reinforced the philosophy of matter derived from creation stories, since materials like bronze were made by mixing purer substances that occurred naturally. Although copper was purer than bronze, it was not regarded as an element, since it appeared to be produced by the earth from other materials. Ancient theories of ore formation often used biological ideas such as birth or plant growth to explain the existence of mineral ores. Many of these theories argued that some combination of earth, water, and heat was responsible for the "growth" of ore. Echoes of these ancient biological ideas can still be heard in our language for mining, as miners talk about rich "veins" of ore or their hope of striking it rich by finding the "mother lode," meaning the birthplace or major deposit of ore.

Mining and metallurgy were some of the most important industries of the early empires. They were so important that often, as in the case of the Egyptians, they came under the direct control of the royal court. In this way, both the production and the use of metal could be controlled by the government. Metal was not, however, the only aspect of material production in ancient times. Glass making was also an ancient development that used the power of fire to transform materials. The oldest remains of glass furnaces are probably those at Tel-El-Amarna, from around 1400 B.C.E. Beads, rods, containers, and figurines of glass date back to this era. The transformation of sand by fire reinforced the concept of the classes of matter, since in some ways sand was the most elemental "earth" possible, and the hotter the fire, the more elemental it seemed to be.

Other metals, such as lead, tin and, later, iron, were also important products. The search for these materials helped to shape empires, as sources of these materials were highly sought after and carefully guarded. For example, part of the reason that the Romans invaded the British Isles was for tin. It was iron, however, that led to another transformation of world history. The ability to mine and smelt iron was one of the most significant technological breakthroughs in human history. Iron appears in the Bible in Genesis, when Tubal-Cain is described as "an instructor of every artificer in brass and iron." Although many world history texts present the Hittites as the people who first smelted iron and then used it to conquer their neighbors, iron smelting predates the Hittite empire. The oldest known iron artifacts date back to around 3000 B.C.E., when there is some evidence that people along the Nile used iron for axe heads. A battle-axe from Ugarit, in Syria, was probably forged between 1450 and 1350 B.C.E. The tomb of Tutankhamen contained chisels made of iron and an iron dagger, made around 1400 B.C.E. These objects were rare and were probably very expensive to make. It was clear that the technology needed to produce them was not completely understood, but the utility of iron was so great that it spurred a great deal of investigation.

When the Hittites began to produce iron tools and weapons, around 1400 B.C.E., they tried to keep their metalworking methods secret, but with the end of their empire in Anatolia (modern Turkey) around 1200 B.C.E., the skills went with the blacksmiths as they dispersed into the new empires that grew up in the region. It was not, however, until around 1000 to 800 B.C.E. that iron began to replaced bronze as the most significant metal for the making of weapons. The search for new metals and the development of the technology to smelt, create alloys, and transform matter worked together with mythic explanations to produce ancient matter theory. By about 500 B.C.E., the ability to control matter had reached a very high level of sophistication in the ancient world, but it would take a new form of inquiry to go beyond creation stories and the skills of the smith to create a broader understanding of the material world.

The development of creation stories that told how the world came to be and the development of technological skills that allowed people to manipulate the environment to make life possible occurred simultaneously. Our ancestors had three great skills that allowed them to overcome the limitations of human frailty. The first was the ability to communicate. People could explain events, describe the world, and pass on knowledge from person to person and from generation to generation. Combining their communication skills with their second great skill, the ability to cooperate, our relatives could work together to reach objectives that individuals alone could never manage. Their third great skill was the ability to manipulate matter in a way that no other creature could. This skill gave them tools, but, in a broader sense, it also gave our ancestors a view of the world based on the transformation of matter from its raw state into a usable state. As the ability to control matter increased, so did the complexity of the societies that used the skills of miners, smelters, and smiths.

2

GREEK MATTER THEORY FROM THALES TO ARISTOTLE

The craftspeople of the great ancient empires in Egypt, Babylonia, Anatolia, India, and China all developed their skills at metalwork and manufacturing to a very high level of sophistication. For example, in China, glazes were added to pottery starting around 1500 B.C.E., increasing the utility and beauty of ceramics, while kilns hot enough to allow glass making were introduced in Mesopotamia and Egypt by 1400 B.C.E. Ancient people developed their techniques for producing and transforming materials by trial and observation and passed that knowledge down from generation to generation, usually within families or by a system of apprenticeships. While the range of products grew, the philosophical aspects of matter remained largely in the realm of theology. Matter theory was closely tied to creation stories, and the process of transforming matter was regarded as magical.

Starting around 600 B.C.E., a group of Greek philosophers began to look for explanations for the structure and function of matter that did not rely upon mystical powers. It is not clear why the Greeks began to separate the natural from the supernatural. Part of the answer may lie in the fact that the Greek gods were depicted as being far more human than the gods of earlier civilizations. While the gods might be capricious, the *kosmos*, or universe, of the Greek world was orderly. This is not to say that the Greeks did not believe in various gods and spirits but rather that they saw the gods as separate from the physical world. Poseidon might be the god of the sea, but water was not filled with a magical spirit. Those Greek natural philosophers who began a search for an understanding of nature that separated the natural from the supernatural started with a firm belief that nature (that is, the material world) both was knowable and followed fixed rules. Although the Greeks did not create science as we know it today, we trace the origin of our modern understanding of the physical world to these first steps away from the mystical explanation for matter.

The Greeks who looked at matter were struggling with two important questions that would continue to be the focus of the study of matter to the present day. First, what were the basic building blocks of matter? Second, how did those original components go together to produce the world we see around us? There were many answers to those questions from many different thinkers, but certain ideas came to dominate natural philosophy, in large measure because the people who presented them produced a compelling system that brought together matter theory, cosmology (the structure of the universe), logic, and observation in one package. Natural philosophy transformed nature from a mysterious and often supernatural realm into an object of study that humans were capable of understanding through the power of the mind and senses.

The Greek philosopher who started this transformation was Thales of Miletus. Information about him is limited, and what we know comes to us from later writers, since no texts by him have survived. He was probably born about 623 B.C.E., in Ionia, which is in modern Turkey. Thales was believed to have traveled widely and likely visited Egypt and studied in Greece. He conceived of the world as a sphere floating in an eternal sea and argued that water was the most fundamental element. It came in three forms: water, earth, and mist. This was an important philosophical insight, since it demonstrated an understanding of the states of matter (solid, liquid, and gas, in modern terms) in addition to different types of matter.

Thales had a number of students or disciples. Two important thinkers who followed Thales were Anaximander (fl. 550 B.C.E.) and Anaximenes (fl. 545 B.C.E.). Anaximander added fire to the list of elements and proposed a cosmology that placed the Earth at the center of three rings of fire. These rings were hidden from sight by mist. Anaximander was said to have argued that animals were generated from wet earth that was heated by the light of the sun, thereby linking the prime elements to life and other objects in the world. In turn, Anaximenes seems to have picked air as the unifying element and built up a system similar to those of Thales and Anaximander. Although information about the specific ideas of the Ionian school is very limited, its adherents were materialists and monists. In other words, they believed that the world was made of inanimate matter and that all apparent matter derived from a single original kind of matter.

The next group to add to the philosophical ferment was the Pythagoreans. This group has been called a religious movement, a political party, and a cult, and it is best remembered for its interest in and its devotion to the idea of the universe being based on numbers. Although it is unclear if there was a specific person named Pythagoras (the Pythagoreans confused the issue by putting his name to their own work), if he lived it was between 585 and 500 B.C.E. The Pythagoreans were not materialists, but their work shaped ideas about the structure of matter because they contributed ideas about order in the universe, based particularly on mathematics and geometric relationships. They also pioneered work on geometric solids, especially what later became known as the five Platonic bodies: the tetrahedron, cube, octahedron, dodecahedron, and icosahedron. These shapes were the three-dimensional versions of flat geometric figures. Because

Figure 2: Geometric solids.

of their perfection of shape, the Pythagoreans thought that these figures were related to the very foundation of matter.

The Pythagoreans saw these shapes in many parts of nature and thought that by understanding geometry, they could gain knowledge about the structure of the world. There are many examples of geometry to be found in nature, such as the hexagonal shape of a honeycomb, the circular motion of the moon around the Earth, and the curl of a snail's shell. The most spectacular examples of three-dimensional geometry found in nature were crystals, which seemed to reveal the hidden geometric reality of the earth. There is a direct link between the ancient philosophers' interest in crystal and our modern understanding of the elements through crystals, since the principles of geometry worked out by the ancients were used by modern chemists when they explained the structure of crystals using the angles at which atoms are joined. In a broader sense, the Pythagoreans contributed the idea that nature was not just orderly but also mathematically precise and certain.

The idea that the elements were uniform and indivisible bodies (as opposed to an undifferentiated mass) can be traced to Leucippus of Miletus, who lived around 420 B.C.E. While the details of Leucippus's life are fragmentary, he was probably the first true atomist, arguing that there were only two aspects of the universe: matter and the void. Matter was made of atoms, which were individually imperceptible, but solid and came in a finite number of shapes. By combining the basic shapes, all common matter could be constructed. Atoms also had the property of motion. This was a necessary aspect of atoms, since the void, being nothing, could not move, so the ability to move had to be inherent or part of the nature of matter.

Following in the atomist tradition was Democritus of Abdera (fl. 370 B.C.E.). Democritus may have been a student of Leucippus, but his ideas expanded from the basic notion of atoms and the void to a complete physics based on the motion and combination of the indivisible particles. In addition to arguing that all matter was made up of atoms, he suggested that the universe functioned as a gigantic machine, with all existence explained by the matter and motion of the indivisible and eternal atoms. In this system, there was no place for gods or the supernatural. In fact, the extension of Democritus's philosophy led him to deny free will, since all of the material world, including humans, was created by the interactions of particles, and those interactions were strictly governed by the inherent nature of the particles. In other words, the universe could not be another way, since the very structure of the universe was determined by the nature of the particles that made up the cosmos. This was a disturbing idea for many, since it seemed to imply that everything was predetermined, leaving no space for the gods or human free will. The idea that the universe was simply the result of the necessary and completely mechanical interaction of particles would appear a number of times as an objection to matter theory.

The Greek philosophers were also greatly concerned with the concept of change. The leading proponent of the idea that the universe was constantly in flux was Heraclitus of Ephesus (c. 544–484 B.C.E.). His great image of change was fire. Fire was always in motion and could transform other things, but fire was also always itself. Another expression of the constant change of nature that has been attributed to Heraclitus was the statement that you cannot step into the same river twice. At the most mundane level, that statement reflects the fact that the particles of water you touched the first time have been swept downstream, so the material of the river has changed. On a more profound level, Heraclitus was also saying that the person taking the action has changed and that each of us is a product of an ever-changing series of experiences.

On the other side of the debate was Parmenides (fl. 480 B.C.E.), who argued that change was logically impossible and that even our concept of motion was an illusion. Change would require things to go from "not being" to "being," and, by definition, things that do not exist cannot contain within themselves anything that exists. Parmenides' most famous disciple was Zeno (fl. 450 B.C.E.), whom we remember because of his explanation of the impossibility of motion, known as Zeno's Paradox. There are various versions of this paradox; in its simplest form, Zeno suggested that a person can never move from point A to point B. The problem is that to get from A to B, the person must cover half the distance to B. We will call this midpoint C. Getting to C takes a certain amount of time. But, wait, says Zeno, to get to C, the person first has to get to the midpoint between A and C. Since there are an infinite number of midpoints possible, and since even the smallest distance takes time to cover, the distance from A to B can never be crossed since it would require infinite time. Motion is thus impossible.

Neither philosophical position came to dominate Greek ideas about change, but it was this kind of debate that helped to make Greek philosophical inquiry so powerful. We still think about the ideas of Heraclitus and Parmenides.

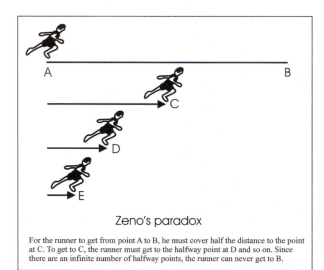

Zeno's paradox

For the runner to get from point A to B, he must cover half the distance to the point at C. To get to C, the runner must get to the halfway point at D and so on. Since there are an infinite number of halfway points, the runner can never get to B.

Figure 3: Zeno's paradox. For the runner to get from point A to point B, he must cover half the distance to the point at C. To get to C, the runner must get to the halfway point at D and so on. Since there are an infinite number of halfway points, the runner can never get to B.

The center of Greek philosophy shifted to Athens around 400 B.C.E., with the rise of Plato and his school, the Academy. Plato (428–347 B.C.E.) was an Athenian from a wealthy, patrician family. He had been the foremost pupil of Socrates (470–399 B.C.E.), but, unlike his teacher, he did not reject the necessity of understanding the physical world. For Socrates, only the Ideal and the nature of man were worthy of study. The Ideal was perfect, mathematical, and divine. Although Plato accepted the superiority of the Ideal over the material, he was also interested in human existence, writing extensively about politics and social organization. It followed that even if the material world might be an imperfect reflection of the Ideal, to make the best of one's life a person needed to understand both the material world and the Ideal. The best presentation of Plato's ideas is found in the *Timaeus,* which includes a description of the structure of the universe.

> If the universal frame had been created a surface only and having no depth, a single mean would have sufficed to bind together itself and the other terms; but now, as the world must be solid, and solid bodies are always compacted not by one mean but by two, God placed water and air in the mean between fire and earth, and made them to have the same proportion so far as was possible (as fire is to air so is air to water, and as air is to water so is water to earth); and thus he bound and put together a visible and tangible heaven. And for these reasons, and out of such elements which are in number four, the body of the world was created, and it was harmonized by proportion.[1]

Plato's material world was divided into realms. The terrestrial realm was imperfect and was based on the four elements of earth, water, air, and fire. These elements were made of particles representing the first four Platonic solids:

Cube = earth
Icosahedron = water
Octahedron = air
Tetrahedron = fire

The celestial realm, being the region from the orbit of the Moon out to the sphere of the stars, was perfect and was composed of a fifth substance, called the quintessence (meaning "fifth essence") or ether. This material was thought to be undifferentiated, meaning that it was more basic than the four terrestrial elements and had not been corrupted or changed since the origin of the universe. Different Greek thinkers argued that the ether was either like a solid, giving us crystalline spheres that contained the stars and planets, or like a superrarified air, so pure and light that it was beyond human sense. The shape of ether was associated with the dodecahedron and was found only in the heavens.

The basic structure of the cosmos was determined by the distribution of matter. Each type of matter had a natural place and always attempted to return to its proper zone. Thus, a rock falls toward the earth when dropped because, being made primarily of earth element, it is trying to reach the lowest zone, while smoke, being primarily fire element, rises up to reach its natural zone above earth, water, and air. The concept of natural place could, in turn, be used to explain many aspects of motion in nature.

This system borrowed from earlier thinkers, such as the Pythagoreans and the Ionians, but Plato rejected the strict materialism of the atomists, and he also rejected the concept of the void. All of the material world must consist of matter, with no spaces between the particles. There was the divine in Plato, as well, both in the form of the Demiurge, a kind of creator deity who was portrayed as the personification of reason, and a belief in a world soul. Plato argued that to understand the world, humans had to have a conception of the ideal forms that underlie all matter and ideas. Plato's theory of forms suggested that the ideal forms must exist separately from the objects of the world and come into the mind from outside. Since people could not be perfect but the ideal (by definition) was perfect, there had to be some way to connect the perfect realm with the imperfect human mind. That link was the divine world soul.

Plato's most famous pupil was Aristotle. Aristotle came from Stageirus, on the Chalcidic peninsula of northern Greece, where his father, Nicomachus, was a physician. Aristotle would probably have followed his father into medicine, since medical education and practice were passed down from father to son, but Nicomachus died when Aristotle was about 10 years old. He was raised by a relative, and, in 367 B.C.E., at the age of 17, Aristotle became a student at the Academy. He stayed for 20 years, first as a student and later as a teacher. When Plato died, Aristotle may have expected to become the head of the Academy in Athens, but the position went to Speusippus, who was Plato's nephew. Little is known about Speusippus; he seems to have followed some of Plato's ideas but rejected Plato's theory of forms. Aristotle left the Academy, partly because of his situation at the Academy and partly because of political turmoil in Athens. He traveled to Macedonia, where he tutored Alexander, son of King Philip. When Alexander became king, he supported Aristotle's creation of the Lyceum, a rival school in Athens.

Aristotle continued and extended the four-element theory of matter. His definition of an element can be found in *De caelo*, where he says:

> An element, we take it, is a body into which other bodies may be analyzed, present in them potentially or in actuality (which of these, is still disputable), and not itself divisible into bodies different in form.[2]

In some senses, this definition sounds modern, since it argues that elements are real matter and cannot be subdivided or reduced to discover a more basic form of matter. It then followed logically that the physical world (the "other bodies") was built up from the elements. Aristotle placed a small caveat in the definition when he talked of composite bodies being made of elements "potentially" or "in actuality," because there was some dispute among philosophers about how elements in composite bodies expressed their characteristics. Despite allowing for the idea that elements might act through potential (referring to that which had the ability to coming into being), he treated matter as if it were really made of elements.

The four elements were strictly terrestrial matter found within the terrestrial sphere bordered by the path of the Moon. Aristotle (and almost all Western natural philosophers prior to Copernicus) set the Earth at the center of the universe and gave it characteristics different from those found in the celestial realm. Aristotle's celestial region was similar to Plato's and was made of a unique substance called the ether or the quintessence. The need to posit the existence of a different element for the celestial realm was tied to motion. On the Earth, all natural motion was seen to be linear (e.g., a rock falls straight down to the ground). While "unnatural" motion (e.g., throwing a rock) could briefly overcome the natural motion of objects, it could not be the basis for understanding nature. Motion in the heavens, such as the path of the stars, sun, and moon, was all circular. Since the natural motion of an object was an inherent characteristic of the matter of the object, celestial and terrestrial elements could not be the same.

In addition to the four elements of earth, water, air, and fire, Aristotle also included the four qualities and the four causes. The four qualities were set in contrasting pairs: hot/cool, wet/dry. At the simplest level, the qualities linked the four elements; the combination of water and air was wet, while fire and earth was dry. There was a deeper purpose to the qualities, however. They were necessary to explain the observed condition of physical objects, since the simple proportions of each element was not sufficient. For example, blood could not simply be described as a particular mixture of water and fire elements, since it was qualitatively different from bile but must have some of the same elemental makeup.

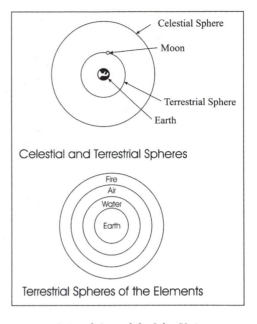

Figure 4: Aristotle's model of the Universe.

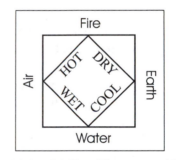

Figure 5: Aristotle's Four Elements and Four Qualities.

Although elements and qualities were useful, they did not provide a complete description of matter, since matter must have a form. That form must be applied to the matter by an active principle (the creator or former of matter) and for a certain purpose, since nothing in Aristotle's system of nature existed for no purpose. To complete any description, these facts must also be accounted for, and that was the role of the four causes. The four causes were formal, efficient, material, and final. The easiest way to understand them is to consider the building of a brick wall. If we look at a wall, we can identify the elements and the qualities by observation, but, to understand what the wall is, we must also recognize its form, origins, and purpose.

The formal cause of the wall was the plan. Plans can exist separate from actual matter, but they are constrained by the limits of matter. We can imagine a 1,000-foot-tall wall that is only one brick thick, but we cannot make such a wall in reality.

The efficient cause is what brings the wall into being. In this case, it is the mason. In the biological world, the mother is the efficient cause of a child, and the Earth is the efficient cause of minerals.

The material cause of the wall is the bricks and mortar. These are the composite materials that are made of the four elements. Since bricks are made almost entirely of earth element, they rest on the ground as their natural zone.

The final cause is the purpose for which the wall is constructed. Our brick wall might have been planned and built to keep the neighbor's goats out of our garden. Just as the wall would not have been planned or built without a purpose, so too, in nature, nothing material exists without a purpose. This system is called "teleological" because it is based on the idea that nature has some end plan or purpose for the existence of objects in the material world. The final cause of things is often not obvious, and it is part of the philosopher's job to figure out the final purpose.

Aristotle's aim was to produce a complete system of knowledge that made clear how all physical aspects of the world functioned. This objective represents both the power and the problem of Aristotle's grand unified system. Aristotle's matter theory allowed for a complete description of physical objects and accounted for both their existence and their behavior. In this way, it was a powerful system, and the basic ideas on which it was based dominated matter theory in the Middle East and Europe until the seventeenth century. In Aristotle's system, natural philosophers had constantly to seek for the final cause of everything from the placement of the nose on the human face (to separate the eyes) to the purpose of clouds in the sky (to cool the earth). This led to a deep examination of matter, but it also led to much fruitless and often misleading speculation about the purpose of things.

Aristotle also commented on the origins of matter, suggesting that in the beginning there was only one element, a prime element, which was differentiated into the four terrestrial elements at the time of the creation of the physical universe. Although Aristotle never suggested that prime matter existed in reality (it had been completely converted when the other elements were created), this hint became the basis for much of Western alchemy. European alchemists in particular believed that through purification, one could isolate prime matter and then transform it into any desired form. This was often associated with the "Philosopher's Stone," and it was the philosopher Aristotle for whom the Philosopher's Stone was named. The exact nature of the Philosopher's Stone was different for different alchemists. Some saw the Stone as an actual physical object that possessed the magical ability to turn matter from one form to another and acted as a kind of catalyst that remained unchanged by the process of transmutation. Other alchemists saw the Stone as a representation of prime matter, so that the new matter came from the Philosopher's Stone itself. For later alchemists, the Philosopher's Stone became a metaphor for the knowledge of transmutation. In other words, there was no physical object called the Philosopher's Stone, but achieving an understanding of the secrets of nature could be talked about as finding the Stone.

Although Aristotle developed a powerful matter theory, he was, like Plato, not an atomist. He argued that there were too many logical problems with the idea of indivisible, imperceptible objects that nonetheless had fixed existence. While Platonic and then later Aristotelian philosophy came to dominate Western thinking about the material world, the atomist thought of Democritus did not disappear completely. The philosophical school founded by Epicurus of Samos (341–270 B.C.E.) based its theory of matter on atoms. Our best record of Epicurean atomism comes from the Roman poet Lucretius (95–55 B.C.E.), who wrote a poem entitled *De rerum natura* (On the Nature of Things). Lucretius says:

> I prove the supreme law of Gods and sky,
> And the primordial germs of things unfold,
> Whence Nature all creates, and multiplies
> And fosters all, and whither she resolves
> Each in the end when each is overthrown.
> This ultimate stock we have devised to name
> Procreant atoms, matter, seeds of things,
> Or primal bodies, as primal to the world.[3]

In other words, atoms are the basic stuff of nature, the primary and most foundational, and all of matter was built up of these "primordial germs." The suggestion that the structure of atoms was hard (since they were indivisible) implied that atoms could not fill all space and there had to be regions that did not contain atoms. This would be like filling a jar with marbles. Where Aristotle and Plato objected to the very idea of a void, the Epicureans were comfortable with the idea, and it provided an easier way to describe motion. Since the void could not oppose motion, atoms were free to move about, while a universe filled with matter suggested that motion was unnecessarily complicated or might even be impossible. Lucretius went on to say:

> Nothing Exists Per Se Except Atoms and the Void
> But, now again to weave the tale begun,
> All nature, then, as self-sustained, consists
> Of twain of things: of bodies and of void
> In which they're set, and where they're moved around.

While the Epicurean model appeals to us as being closer to the modern view of atoms and is sometimes pointed to as the origin of atomic theory, in historical terms, it had little effect on the development of natural philosophy. The reason for this was twofold. First, the Epicureans were known less for their physical theories than for their ideas about social and personal behavior, favoring the pursuit of pleasure. Our modern use of the term "epicure" for someone who enjoys good food and drink reflects this. The second reason that the Epicureans had little influence was that their work was not well received by the intellectual communities that developed after the decline of Greece and Rome. Christian, Hebrew, and Islamic scholars tended to regard the Epicureans as hedonistic and viewed their work with suspicion or even hostility, and thus little effort was made to investigate or revive Epicurean philosophy.

Greek philosophy, particularly that of Plato and then Aristotle, came to dominate philosophy throughout Europe and the Middle East even after the end of the age of Greece and Rome and the closing of the Academy and the Lyceum. It was critiqued, modified to fit with accepted theology, and absorbed by Christian, Hebrew, and Islamic scholars. In its new form, Greek philosophy became part of higher education. Aristotle's matter theory, because it accorded well with the cosmology of the new scholars, remained largely unchanged. This did not mean that it was accepted without question, but medieval scholars who studied natural philosophy adopted the general principles of the Aristotelian system, even if they challenged certain of the great philosopher's individual observations.

Aristotle's matter theory did what it was expected to do: explain the structure of the cosmos and the observed action of matter. What it did not do was look for quantitative information about nature. While Aristotle assumed that a tree or a person was made up of a particular proportion of each of the four elements, he did not attempt to determine what the exact mix was or how altering the proportions could produce something new or different. Aristotle's philosophy was based on careful observation and logic, not experimentation. Aristotle objected to the idea of testing nature, since such a test would be, by definition, unnatural. The direct examination of matter was carried out by a different group of people, some of whom were scholars but many of whom had no formal education in philosophy. They were the artisans, smiths, dyers, tanners, glassmakers, and others who would be drawn to matter theory by alchemy and by the prospect of transforming one type of matter into another, and especially by the possibility of turning base metals into precious metal or, in more direct language, lead into gold.

NOTES

1. Plato, *Timaeus*, Bk. I, trans. Benjamin Jowett, www.gutenberg.org/etext/1572.
2. Aristotle, *De caelo*, Bk. 3, Sec. 3, trans. J. L. Stocks, etext.lib.virginia.edu.
3. Lucretius, *On the Nature of Things*, trans. William Ellery Leonard, classics.mit.edu/Carus/nature_things.mb.txt.

3

MATTER THEORY IN CHINA, THE MIDDLE EAST, AND INDIA

The pursuit of knowledge about chemistry depended on a combination of theory and practical investigation. While the early philosophers supplied theory, it was others who over time supplied practical knowledge about matter. Some of that knowledge came from the everyday work of smiths, potters, brewers, and other artisans, but the hunt for the secrets of matter came from people we now collectively refer to as alchemists. The term "alchemist" generally refers to a person who attempts to transform matter from one form to another. Although some alchemists were seeking to create precious metals (silver and gold) from nonprecious metals (also called base metals) like lead and mercury, others sought to transform the nature of matter in a medical sense, so as to gain immortality or cure disease. The most scholarly or esoteric study of alchemy sought to find the secret or hidden rules of the universe and could be as much a spiritual quest as a practical investigation of matter. Some alchemists, however, were con artists, seeking to profit from the public's belief in transmutation and from their greed. The distinction between charlatan and scholar was not always clear, since a number of alchemists who were genuinely seeking the secrets of matter resorted to trickery to gain patronage support for their true research.

Alchemists came from all parts of society, from the lowest peasants to monarchs and popes. While it would be an error to think of alchemy as "scientific" in the modern sense of the term, it was decidedly more hands-on and experimental than the philosophical ideas that had been developed by Greek philosophers such as Plato and Aristotle. In historical terms, it is interesting to note the similarity of interest in the transmutation of matter found in Europe, the Middle East, India, and China. Historians have long speculated about the contacts between East and West and whether alchemical ideas grew up independently in different places or were imported. Although it was formerly believed that there was very limited contact between Europe and China prior to the circumnavigation of the globe by

European explorers, more recent archeological evidence suggests that trade and travel were more extensive in the ancient world than was previously thought. In particular, the rise and spread of Islam after 800 C.E. led to an extensive network of trade and information exchange that reached as far as Indonesia in the east and Spain in the west.

The alchemical study of matter was not a uniform study, and, given the diverse range of people who practiced alchemy, it is not surprising that there were wild variations in what people were doing and in their ideas about what they saw happening. In addition, alchemists were by nature inclined to secrecy, so what we know of their work is fragmented and incomplete.

CHINA

In China, much of the alchemical tradition is tied to Taoism, both as a spiritual pursuit of understanding and because important Taoist figures practiced alchemy. The origin of Taoism is attributed to Lao-tzu (the Old One), who was supposed to have lived in the sixth century B.C.E., but there is little direct evidence for this. It seems more likely that in the later Zhou period, between 600 and 221 B.C.E., there developed a philosophical school that looked to nature (Tao) for inspiration about the proper way to live. This had both practical and mystical aspects and was directed at guiding people to a happy and ordered life. The primary document of Taoism is the *Tao te Ching* (Book of the Way) which was written in the third or perhaps fourth century B.C.E. The great model for Taoist thought was water, which conformed without losing its nature, flowed around obstructions, and was always part of nature.

The Taoist interest in nature led to a great deal of study and produced strong resources in medicine and practical arts such as smelting. Alchemy was part of the activities of the Mao Shan school, where an important synthesis of Taoism thought took place. Taoism was not an easy philosophy to understand, however. Taoist writing was frequently obscure and full of riddles. One of the most famous sayings of Taoism was that "those who know, do not talk; and those that talk, do not know." By the logic of this dictum, anyone who taught Taoism or any text that tried to explain it would be untrustworthy. Thus, wisdom and knowledge were highly prized, but sharing them was a tricky and potentially dangerous proposition. This matched perfectly the concerns of the alchemist.

Chinese alchemists in the Taoist tradition were interested in transmutation, but more frequently they sought enlightenment and immortality, rather than material wealth. Taoist alchemists thus often more inclined to the medical or pharmacological end of the spectrum, producing tinctures, potions, and pills and devising alchemical regimes to aid in the pursuit of immortality.

The interest in alchemy did not exclude philosophical or theoretical aspects of the study of matter, and a number of different systems of matter theory were developed in China. A four-element system based on earth, water, air (or wind), and fire can be traced back to Chinese creation stories, but a more popular system, based on five elements, was well established by the time of the introduction of Taoism, and its roots are lost in history. The five elements of

	Wood	Fire	Metal	Earth	Water
Color	Blue/green	Red	Yellow	White	Black
Symbol	Dragon	Phoenix	Caldron	Tiger	Tortoise
Season	Spring	Summer	Transitional	Autumn	Winter
Direction	East	South	Center	West	North
Planet	Jupiter	Mars	Saturn	Venus	Mercury
Sense	Sight	Taste	Touch	Smell	Hearing

classical Chinese thinking were wood, fire, metal, earth, and water. These had a long series of associated factors, such as color, direction, and planet.

The Chinese system shared characteristics with the Aristotelian system in that it provided a way of describing and ordering the world. Unlike the Greek system, the Chinese system also attempted to integrate the natural and the supernatural worlds. As such, Chinese astrology, Taoist thought, and aspects of Confucianism and Buddhism became bound up in matter theory. The deep symbolism of both the planets and the figures associated with the elements increased the importance of knowledge about the material world, because such knowledge opened the door to knowledge of the far vaster supernatural world. The five-element system was also less strictly material, since the term "element" in the system could be replaced with the term "fundamental material" in the sense that the Chinese did not see all of the elements as indivisible. Wood and metal, in particular, were not a single type of matter but a class of matter.

As well as the five elements, the philosophy of Yin and Yang influenced Chinese philosophy and matter theory. Yin and Yang represent linked opposites, or antagonistic factors. Such dualities as male/female, water/fire, and night/day are antagonistic pairs, but the parts of the pairs cannot exist without their opposites. In some versions of this philosophy, the antagonists parts must contain something of their opposite, giving rise to the most famous of the Yin Yang symbols. The functioning of nature was based on balance, and the objective of the wise was to reach a state of balance intellectually, spiritually, and socially.

Figure 6: Yin Yang symbol.

Matter theory, alchemy, and the skills of the artisan mixed together in various ways. For example, Chinese metalworkers introduced the double-acting bellows for forge work around 310 B.C.E., increasing both the efficiency and the heat of their furnaces. Around 300 B.C.E., they introduced cast iron; by combining various technologies, they produced the first water wheel–powered bellows for iron smelting by 31 C.E. The forge and furnace took on mystical qualities, and deities and spirits were invoked while smiths, and alchemists, did their work. Many of the most important early metallurgical projects were the casting of temple bells and statues. Such projects were accompanied by great ceremony and required spiritual effort, not just physical work. There could be great rewards for those who created wondrous objects.

One of the earliest direct references to alchemical activity was the imperial edict that ordered the public execution of those caught making counterfeit gold. The edict, supposedly issued in 144 B.C.E. by the Emperor Jing, was designed to curtail robberies committed by false coiners, who, after spending much time and money on their efforts, were left with nothing and turned to banditry. The ban does not seem to have lessened interest in alchemy, however, since, in 133 B.C.E., the Emperor Wu brought an alchemist to the imperial court. The alchemist offered Wu a way to gain immortality, partly through the creation of alchemical gold and partly through worship of the goddess of the stove. The alchemist first made gold vessels for the emperor to use for his meals; through appropriate offerings and rituals, the emperor would please the goddess of the stove, who would then grant the emperor an audience with the immortals of Penglai. If the emperor followed all the steps correctly, the immortals would grant him the secret of immortality.

One of the most famous stories of early Chinese alchemy related the work of Wei Po-Yang. The story was only partly about alchemy, being as much a morality tale about proper behavior.

> Wei Po-Yang entered the mountains to make efficacious medicines. With him were three disciplines, two of whom he thought were lacking in complete faith. When the medicine was made, he tested them. He said, "The gold medicine is made but it ought first to be tested on the dog. If no harm comes to the dog, we may then take it ourselves; but if the dog dies of it, we ought not to take it."

Po-Yang gives his secret medicine to the dog, and instantly the dog dies. This seems to indicate that Po-Yang has either not completed the process or made some error. This places his disciples in a quandary, since they believe that he has the secret knowledge and it would be difficult to criticize the master. They ask Po-Yang if he would take his own medicine, but his answer does not really clear up the disciples' uncertainty.

> I have abandoned the worldly route and forsaken my home to come here. I should be ashamed to return if I could not attain the *hsien*. So, to live without taking the medicine would be just the same as to die of the medicine. I must take it.

When Po-Yang takes the medicine, he appears to die in the same manner as the dog. One of the disciples decides that Po-Yang must know what he is doing, while the other two abandon the whole project. This is the crucial point of the story, which is about faith, both in the master and in the ways of nature.

> "Our teacher was no common person. He must have done that with especial intention." The disciple also took the medicine and died. The other two disciples said to one another, "The purpose of making medicine is to attempt at attaining

longevity. Now the taking of this medicine has caused deaths. It would be better not to take this medicine and so be able to live a few decades longer."[1]

Po-Yang revives and wakes up the faithful disciple and the dog, and they go on to join the immortals. The two disciples who lost faith are informed that their ex-master has transcended mortality, and they must live a life full of regret over their missed opportunity and lack of faith. The story is full of symbolism and the importance of faith and obedience. The alchemist (or those who would follow the alchemist's path) had to understand not only the practical issues of making tinctures or medicines but the deeper esoteric or hidden aspects of nature. What makes the story even more complex is that the central character, Po-Yang, is a thinly disguised Lao-tse, founder of Taoism, one of whose names was Po-Yang of Wei. Po-Yang is thus not merely a master alchemist but a spiritual master.

Chinese alchemy influenced the study of matter both directly and indirectly. Directly, the Chinese alchemists studied a huge range of materials and engaged in investigations of the properties of plants, animals, and minerals. The discovery of gunpowder, for example, was almost certainly the work of an alchemist. The earliest clear mention of gunpowder comes from Zeng Gongliang, who in 1044 compiled the volume *Collection of the Most Important Military Techniques*, although gunpowder was probably known before this date.

Indirectly, the material wealth of the Chinese empire stirred the imagination and the greed of people in other parts of the world. Trade routes were also paths for the exchange of information, but many of the most important secrets, such as the formula for porcelain or the secret of silk, were well kept by the Chinese. This stimulated both exploration and investigation as outsiders sought to duplicate the treasures of China or create their own gold to purchase what China had to offer.

Between the secrecy of the alchemists and the lack of surviving historical evidence, it is not clear to what extent alchemical influences flowed from East to West and back again. What is clear is that western alchemists, particularly European alchemists, believed that the Chinese knew important secrets about the manipulation of matter. The existence of the unknown but fabled land of Cathay was a spur for alchemical studies elsewhere. For the charlatans, the expensive and magical materials used to turn lead into gold had to be imported from far away China. It was a perfect excuse to charge huge prices for nothing.

MIDDLE EAST

The English word alchemy comes from the Arabic word *al-kimiya'*, and a wide range of alchemical and chemical ideas come from the work of Arabic and Islamic scholars. Islamic alchemy differs from that of China in a number of important ways. The first is historical. The roots of Chinese alchemy are lost in the mists of history, and the Chinese tradition of attribution to historical figures further clouds the story. Islamic scholars, although often secretive, tended to present their work as their own. Islamic scholars also drew on Greek, Roman, Christian, and Hebrew sources. These texts were scattered through the remnants of the Roman Empire that were taken over by various Islamic empires that expanded from the Arabian peninsula and took over north Africa, parts of Europe, and the eastern Mediterranean region. Later, when the Middle East became a trade cross road between Europe, Africa and the Far East, especially China, Islamic scholars also encountered ideas about matter from India and China.

Another difference was religious. Islam's powerful monotheism was much different from the broad mytho-poetic pantheon of Chinese religion, or the philosophy of Taoism. It followed that it held quite different views on the nature and function of the supernatural. The realms that were open to human investigation were more clearly delineated in Islam than in other religions of the region; essentially, Islam held that all of nature was open to study. Although there were some aspects of natural philosophy (such as botany) that were hard to investigate because of the prohibition on the making images of God's creation, the study of matter was very popular.

What the Islamic scholars shared with their Chinese and Indian counterparts was a strong practical aptitude. Where the Greek philosophers often regarded the work of the artisan and the trader as being of very low status, Islamic scholars were expected to have practical skills. Many of the most important Islamic natural philosophers were trained as physicians. Gentlemen and *hakim* (wise or educated men) were expected to both understand the world and to participate in it.

Although many of texts from early Islamic alchemical studies have been lost over the years, Islamic alchemical tradition says that the first Islamic leader to be interested in alchemy was Khalid ibn Yazid, a Umayyad prince of the seventh century C.E. He was said to have been taught alchemy by a Byzantine monk named Maryanos or Morienus and, as part of his interest in the subject, ordered manuscripts translated from Greek and Coptic into Arabic. Although a manuscript entitled *Correspondence of Marianos the Monk with the Prince Khalid ibn Yazid* exists, its authenticity is uncertain, and no records independently indicate that Khalid ibn Yazid wrote it or was even interested in alchemy. His name may have been used to legitimize the study of alchemy in the Islamic world.

A more certain and important foundation for Islamic alchemy and matter theory was the work of Jabir ibn Hayyan, also known as al-Harrani and al-Sufi. There is much scholarly debate over the extent of Jabir's work and whether he even existed as a single historical figure. The traditional biography of Jabir sets his birth at around 721 C.E., and his death at 802 C.E. or even as late as 815 C.E. This was just at the beginning of what historians have called the Islamic Renaissance, a period of great intellectual and artistic activity. Jabir was the son of a pharmacist who lived in Kufa, but his father was involved in political activity against the caliph and was executed. When Jabir was old enough, he was sent to study in Arabia. He then worked as a physician, likely in Baghdad, and was under the patronage of the Barmaki Vizir during the Abbssid caliphate of Haroon al-Rashid.

Like that of many Islamic scholars, Jabir's work was not really confined to a single topic. Over the generations, more than 1,500 texts have been attributed to him, and they range in subject from astronomy to religious history. What seems most likely is that there was an actual Jabir and that he wrote a number of influential texts, but over time others have modified and added to the collection, clouding the historical record and perhaps supplanting the original texts.

Jabir developed an elemental theory that resembled the Aristotelian system, but he transformed the foundation from the material elements to the qualities. He believed that matter was ultimately composed of four "natures": hot, cold, moist, and dry. Unlike the familiar Aristotelian qualities, Jabir's natures were not abstractions or additions to matter. He presented them as real, material, and independently existing entities. By combining the natures, material elements (or second elements) were created. In other words, the material world was made of compounds, whereas the natures were immutable. It followed that those compounds could be transformed in various ways by manipulating the balance of natures. This contributed to the idea that alchemical action was seeking beneath the material for the eternal and that the transformation of material was linked to the proportions of the underlying natures.

Another aspect of Jabir's work was the separation of supernatural and human creation. God had created the world by creating the natures, but since the material elements (fire, air, water, and earth) were secondary, they lay in the purview of the worldly and therefore could be "created" by humans. It thus would not be a transgression against God to do alchemy, so long as it was the elements and not the natures that the alchemist was creating. With this system, Jabir went on to explain one of the most perplexing problems in matter theory regarding the formation of metals. Metals and metal ores were found primarily as veins in different kinds of rock. This raised questions about the formation of metals, since they seemed to have been produced from unrelated matter. Jabir argued that all metals were composed of sulfur and mercury in various proportions. Since sulfur and mercury contained the natures in the state closest to the pure, they offered a path to the construction of metal.

In addition to the theoretical end of alchemy, Jabir contributed to the practical study of matter. For example, sal ammoniac was introduced to natural philosophical studies by Jabir (or a pseudo-Jabir). Two varieties of sal ammoniac were distinguished: natural (ammonium chloride), called *al-hajar,* and derived (ammonium carbonate), referred to as *al-mustanbat.* The latter was created by the dry distillation of hair and other animal materials. The invention of the alembic (a distillation vessel) is attributed to Jabir, along with dozens of chemical procedures, such as double distillation and the use of sand baths.

Another of the great Islamic alchemists was Abu Bakr Muhammad ibn Zakariyya, better known as Al-Razi. His birth date is uncertain, with estimates ranging from 825 to 854 C.E., but he is thought to have died in his home city of Rayy in 925 C.E. Al-Razi was interested in a broad range of topics and worked primarily as a physician. As such, his chemical work often reflects an interest in pharmacological aspects of matter. Al-Razi's philosophy differed from the Aristotelian and Platonic heritage in that he developed an atomistic model of matter, arguing that bodies (that is, material objects) were composed of indivisible elements and the space between them. The space, or separation between particles, affected the characteristics of matter. Characteristics such as softness and hardness or transparency could be accounted for by the density or amount of space in the elements involved.

The best-known work of alchemy by Al-Razi is *The Book of the Secret of Secrets*. Although the title suggests mysterious and arcane knowledge, it is far more like a guide to setting up and running a laboratory than the spellbook of a wizard. Al-Razi does discuss aspects of transmutation and various elixirs, but much of the text is composed of lists of materials and equipment, many of which would be perfectly familiar to modern chemists, such as glass beakers, funnels, water-baths, and tongs. He also collected, listed, and described a large range of specimens of minerals, such as lapis lazuli, gypsum, borax cinnabar, lime, potash, and the naturally occurring metals. This kind of geochemical collecting has a long history in chemistry, and it would become a significant aspect of the hunt for new elements in a later era, as chemists analyzed the properties of rare minerals to discover previously unknown substances.

One of the most important products of Al-Razi's work was his classification system. Although the four elements that had developed from earlier roots remained at the heart of the theory of matter, Al-Razi's interest in practical application led him to create a broader method of grouping materials, in large

Al-Razi's Table of Chemical Substances

Mineral	Vegetable	Animal	Derivative
	(little used)	Hair	Litharge (lead oxide)
		Skull	Red lead (tin oxide)
		Brains	Verdigris
		Bile	(acetate of copper)
		Blood	Tutia (zinc oxide)
		Milk	Cinnabar
		Urine	(mercuric sulfide)
		Eggs	Glass
		Mother of pearl	Caustic soda
		Horn	Various alloys
		etc.	etc.

part by their characteristics. This showed clear attention to the real world and evidence of experimentation, not just observation and theory.

Islamic scholars preserved Greek philosophy and added to it a greater interest in practical alchemy. They contributed tools and methods of investigation, as well as discovering and producing new chemicals. In a broader sense, the *hakim* and the centers of scholarship that existed in the Islamic world were copied in Europe as part of the educational system, helping to preserve and promote a love of learning.

INDIA

Western scholars have only recently started to explore the history of alchemy in India. Part of the difficulty of studying this subject comes from the complexity of languages and the diversity of people in what is modern India, Pakistan, and the Himalayan region between the Middle East and China. Yet, it is becoming apparent that work on the structure of the material world was well established by the time of the rise of Greece and, later, the Chinese and Islamic alchemists. Matter theory was even more closely linked to the mytho-poetic and religious traditions in this region than in other parts of the world, so that there was never a clear distinction between the spiritual and the material aspects of the world. Thus, Indian alchemy did not develop as an independent study but was part of a broader investigation of the universe.

The foundation of natural philosophy in India dates back to around the fifth century B.C.E., and the concept of the transmutation of matter can be found in parts of the ancient Vedic writings. The oldest description of elements was based on a five-element system that comprised earth, water, fire, wind (or air), and space. There is evidence of a kind of atomistic view of matter as indivisible particles and even an estimation of their size. According to the ancient text *Tarkasamgrahadeepika,* the smallest perceptible object was a dust mote seen in a beam of sunlight. The atom was then equal to one-sixth of the mote.[2] This atomism was overlaid with a form of vitalism to account for the difference between animate and inanimate matter. Vitalism, which appeared in the alchemy of almost all regions, was the belief in a spirit or spark of life that occurred in some types of matter. If matter contained the vital spark, it was alive. Whether the vital spark could be created or made to appear in inanimate matter was one of the areas of research for alchemists.

There is also evidence of practical interest in matter such as Kautilya's *Arthasastra,* which described the production of salt from the sea and the collection of shells, diamonds, pearls, and corals. There is also archeological evidence of alcohol production using the "Gandhara stills," which date to between 150 B.C.E. and 150 C.E. These were possibly the earliest alcohol stills and certainly provide the earliest evidence of mass production of distilled alcohol. The production of alcohol was an example of the transformation of matter, and the properties of alcohol, such as its flammability, seemed to suggest that it was closer to its basic nature than the material from which it was made. There was a great deal of interest in the medicinal properties of substances and a long tradition of gold- and mercury-based medicines.

Some time between 1100 and 1300, a new mystical system appeared in India. This was Tantrism, a complex blend of naturalism and Hindu and Buddhist ideas. A number of Tantric authors wrote on alchemical topics. By that time, contact with both eastern and western sources of alchemical knowledge was well established, and it is likely that a greater interest in the transmutation of matter and in the search for the elixir of life evident in these outside sources influenced the direction of Tantric studies in alchemy. An early alchemical text by an unknown author from the twelfth century was the *Rasarnava* (Treatise on Metallic Preparations).

One of the most important aspects of Indian alchemy was the use of mercury to create elixirs to cure diseases or prolong life. The preparation methods were complicated, requiring as many as 18 separate operations, known as the *samaskaras*, that included heating, grinding, distillation, mixing with other substances (amalgamation), and liquefaction. At the end, a mercuric tincture was produced and consumed.

The history of alchemy around the world is an expanding area of research for scholars. The three great regions for alchemical studies shared a belief in the possibility of the transformation of matter, but each region placed alchemy in a somewhat different context. The Islamic tradition was very practical, with a strong streak of Aristotelian philosophy. The Chinese also had a strong practical interest in matter but were more interested in the medical and spiritual aspects of alchemy. The Indian tradition tended to place alchemy within the Vedic tradition and used it as a way to understand the universe, rather than as the basis for an independent study of matter. What the history of alchemy has revealed is a constant fascination with the materials of the world and a desire to control matter. It also shows us that there has been a long-term effort to classify and categorize the components of nature and an equally long-standing belief that understanding the basics of matter can connect us with something beyond the material world. When the work of the Islamic and Chinese scholars, along with the revived works of the Greeks, came together in the European medieval world, it started a chain of events that led to the creation of modern chemistry. Although these ancient scholars were interested in the transmutation of matter, the Philosopher's Stone, and the secret of eternal life, these formed only part of a much broader study. In the European era, the pursuit of transmutation became the central focus of alchemy.

Along the way, the concept of the element became more codified, but, at the same time, the emphasis among Chinese and Islamic practitioners on practical investigation meant that alchemy had to be an active study. The apparatus created by people like Al-Razi opened the door to a more experimental approach to matter, and, in turn, the results of those studies began to disrupt the Aristotelian model of the four elements. Although the four elements of earth, water, air, and fire were philosophically elegant, they did not stand up to the challenge of practical application.

NOTES

1. W. W. Benton, "The Elixir of Life," *Hexagon of Alpha Chi Sigma* 20, no. 2 (1938): 2.

2. This was an interesting measurement; the size of the smallest object visible to the naked eye this way is about 40,000 nanometers, so one-sixth of that would be around 6,700 nanometers. While this is a long way from the size of a single atom of hydrogen (0.12 nanometers), it is in the range of large molecules.

4

ALCHEMY AND THE CONCEPT OF ELEMENTS

EARLY WESTERN ALCHEMY

Alchemy developed independently in many places and became a highly sophisticated study in India and China and throughout the Islamic world. The emergence of alchemical studies in Europe, especially in western Europe, where the language of scholarship was Latin, depended on the rediscovery of ancient philosophy and the introduction of alchemical ideas, practices, and tools from the Middle East and Asia. While alchemy in western Europe shared many roots with eastern alchemy and western alchemists worked on many of the same projects as their peers in the East, Europeans did not see alchemy in the same philosophical and religious context. Part of the reason for this was the position of the Church. The Christian attitude toward alchemy was always ambivalent; on one hand, there was a clear link between transubstantiation, where the communion wine and wafer were transformed into the body and blood of Christ, and the concept of transmutation, the changing of one substance into another, while on the other hand, the Church had long-standing prohibitions against sorcery, and alchemy was often closely associated with spirits and the supernatural.

The lack of a central theory or school of alchemy in Europe and the desire of many alchemists to avoid the charge of sorcery by making alchemy less magical contributed to a change in the objectives and style of study. Although alchemy should not be seen as simply a precursor to chemistry, it was in Europe that chemistry emerged as the experimental and material-based study of matter. Questions explored by the alchemists could not be resolved without transforming the study of matter, particularly the elements, from a process that depended on esoteric (hidden) or supernatural forces to one based on natural causes. In a sense, the alchemists researched themselves out of existence, since the theoretical foundation of alchemy was based on transmutation, and the more that was learned about the material world through alchemical studies, the less likely transmutation seemed. The theories of the alchemists became incompatible with the knowledge and methods necessary to expanded knowledge of the material world.

To understand the development of European alchemy, it is necessary to go back and look at the creation of the medieval world. The collapse of the Roman Empire took place over more than a century, with the western half of the empire falling to successive barbarian invasions through the fifth century, while the eastern part of the empire became the Byzantine Empire and survived until the fall of Constantinople to Mehmet II in 1453. With the end of Roman control, western Europe entered a period of turmoil, sometimes called the Dark Ages because the light of learning was lost. While we now know that the period from 450 to 800 C.E. was not as dark as it was once thought to have been, it was a period when intellectual life was curtailed. Many of the scholars who lived at the end of the empire died without transmitting their knowledge to the next generation, schools were destroyed or closed, and tens of thousands of manuscripts were lost. The only organization that was able to preserve any intellectual life was the Church. Church officials saved what could be saved and put some of the ancient knowledge to use when they could.

The most significant aspect of Greek natural philosophy to survive in Europe was medical knowledge, particularly the works of Galen (129–199 C.E.). Galen had been the personal physician to four emperors, most notably Marcus Aurelius. He had investigated anatomy, physiology, and pharmacology, writing extensively on these subjects, as well as on aspects of philosophy. Because of Galen's fame and importance, his work was widely distributed, and many Galenic or Galenic-based texts survived the collapse of Rome.[1] Galen was an Aristotelian, and his medical work was based on a belief in the Aristotelian elements, qualities, and causes. The hints of a greater system in Galen's medical texts helped to keep alive an interest in Greek thought.

The turning point that would help revive intellectual life and interest in the elements was the rise of Charlemagne (c.768–814 C.E.). Charlemagne (also known as Charles the Great) conquered a large portion of Europe, including all of modern France and Germany and most of Italy. In addition to achieving military conquest, Charlemagne attempted to reform the legal system, create civil government, promote the papacy, and modernize education. Charlemagne invited Alcuin (c.735–804 C.E.), a Northumbrian scholar and monk, to become his minister of education. Alcuin oversaw the creation of the cathedral schools, many of which were the forerunners of our modern universities. He increased the literacy levels of the clergy and helped give the small educated class of Europe a stronger place in civil society.

One of the consequences of the Carolingian Renaissance, as it is sometimes called, was a greater interested in regaining ancient knowledge. Christian scholars traveled to centers of learning in Spain, such as Toledo, and to the cities of Alexandria, Jerusalem, and even Baghdad to study and translate Arabic and Greek material into Latin. Among the early Greek works to be translated was

Plato's *Timaeus,* which gave medieval scholars an introduction to matter theory. Known to the Latin scholars as Geber, Jabir ibn Hayyan authored alchemical works that were translated from Arabic during the twelfth and thirteenth centuries, as were the medical and alchemical works of Al-Razi (or Rhazes, in Europe). Commentaries on Aristotle by Avicenna, Alpetragius, and Averroës led to a huge interest in Aristotle's actual work, so that important works like the *Physics, Meteorologica,* and *De Animalibus* were found and translated into Latin by the end of the thirteenth century.

In this burst of translation, the work of Robert of Chester (fl. 1141–1150) stands out in the history of chemistry. Robert of Chester (or Rétines), along with his friend Hermann the Dalmatian, had been asked by Peter the Venerable, Abbot of Cluny, to translate the Q'ran into Latin. Robert then went on to translate the *Book of the Composition of Alchemy,* which he said he completed on February 11, 1144. Robert said of this work: "Since what Alchymia is, and what its composition is, your Latin World does not yet know, I will explain in the present book." The text was attributed to Khalid ibn Yazid, although it is almost certainly not by Yazid and it is not clear who was the original author. Although alchemical ideas may have existed in some attenuated forms in Europe before this date, it was largely as a result of the translation of this text that alchemy came to Europe.

The Arabic influence on alchemy and chemistry can be seen in the nomenclature that came with the sources, and many or our modern terms came from Latin translations of Arabic words.

Arabic origin	Modern term
Al-qali	alkali
Al-kimia	alchemy
Al-kuhl	alcohol
Qarabah	carboy
Al-iksir	elixir
Jargon	jargon (originally a kind of zircon)
Naft	naphtha

The European interest in both the material and the intellectual wealth of the Islamic world was heightened by the military clashes of the Crusades. The first three Crusades (1096–1099, 1147–1149, 1189–1192) were military successes to varying degrees for the Christian forces and led to the creation of Crusader kingdoms in the Middle East along the eastern coast of the Mediterranean. Because of the kingdoms and the connected trade routes overland through Constantinople and by sea to Genoa, trade goods made their way to Europe. Silk, porcelain, spices, glass, metalwork, soap, perfume, and textiles intrigued Europeans and increased interest in trade and exploration.

By the beginning of the thirteenth century, Europe was primed for alchemy. In addition to the intellectual material from the scholars who were rediscovering Greek matter theory and the seemingly mysterious and even magical material wealth of the East, one further condition made Europe receptive to the spread of alchemy. This was the existence of Christian liturgy. In the clearest terms possible, the Church reinforced the idea that alchemy could work. In transubstantiation, the wine and the wafer of the Holy Communion were transformed into the body and blood of Christ. Although there were deep theological arguments about what and how transubstantiation worked, this unintentional theological endorsement of transmutation became one of the basic principles of European alchemy.

One of the most popular manuscripts that introduced the elements to the medieval world was Bartholomew the Englishman's (fl. 1240–1250) *On the Properties of Things*, written sometime after 1230. Part of the reason that this text was so important was that it was translated from the original Latin into French, Dutch, Spanish, and English, thereby greatly increasing its audience. *On the Properties of Things* was largely a compendium of knowledge starting with God, angels, and humanity and working down through the hierarchy of being. Although Bartholomew cited Plato's *Timaeus,* his description of the material world was based largely on Aristotelian material. Bartholomew introduced the elements and qualities in Book 4, saying:

> Elements being four, and so also four qualities of elements of which every body that hath a soul is compounded and made as matter. . . . Man's body is made of four elements: earth, water, fire, and air; and every thereof hath proper qualities. Four there are including the first and principle qualities, that is to wit heat, cold, dry and wetness.[2]

After writing many pages that describe the parts and function of the parts of the body, Bartholomew turns, in Book 8, to the material of the world and the heavens. It is interesting to note in passing that Bartholomew takes as understood that the world is a sphere "contained in the roundness of heaven."[3] He turns to the matter of the world in Book 16, giving a long list of metals and precious stones. For the more important metals, such as gold, silver, and iron, Bartholomew gives the composition, with quicksilver (mercury) representing water (wet and cold) and brimstone (sulfur) representing earth (hot and dry). He describes the working of gold, its heaviness and reflective properties, then explains that gold can be used to prevent swooning, ease the pain of sore limbs, and fight "evils of the spleen."[4]

What is interesting in Bartholomew's work is that there is no indication that matter can be transformed. In this, he is following the matter theory of Aristotle and of many of the Arabic commentators whose ideas form the basis of his encyclopedia. There are some mythic and quasi-magical aspects to his descriptions; for example, gems almost all come from far-off lands, particularly India, where a number of the rarest stones are mined by the Troglodytes. Many of the things discussed have medicinal properties, but no fantastic claims of cures are to be found in *On the Properties of Things*. Despite the lack of alchemical content, Bartholomew's work helped to promote interest in matter theory in two ways. First, *On the Properties of Things* was a major reiteration of Greek and, in particular, Aristotelian matter theory. Second, Bartholomew's work helped promote the study of the material world. Although his work was not directly alchemical, Bartholomew placed a great deal of emphasis on the composition of matter based on the proportion of the qualities as represented by two specific substances—quicksilver and brimstone. It did not take much reading between the lines to reach the conclusion that if you could change the balance of quicksilver and brimstone in something, you could change its qualities and thus produce a new substance.

Although Bartholomew's work was widely read, his fame was far surpassed by that of Albertus Magnus (c.1200–1280), who wrote on everything from mineralogy to theology. He was of upper-class birth and was well educated. He joined the Dominican order in 1223 and held various high Church offices, serving as the bishop of Ratisbon until his retirement, in 1262. Albertus also taught at a number of schools, and it was from his teaching and writing that he earned the nickname "Doctor Universalis," or master of all subjects. Like Bartholomew's, Albertus's works were often compendiums, such as his *De mineralibus* (c.1264), but he had a much greater breadth of knowledge, and much of his nontheological work reflects an interest in direct observation rather than simply being a summary of the work of other authors.

Albertus was aware of the growing interest in alchemy in his age, but he was skeptical about the possibility of true transmutation. He argued that alchemists simply created the appearance of precious metal, not the true metal. Yet, at the same time, he recounted stories such as the one about the power of a toad's gaze to crack fake emeralds and the one about the existence of a type of stone that would allow a person to understand the language of birds. In terms of matter theory and the elements, he was a devoted Aristotelian. Although he did not think Aristotle infallible, he accepted the basic element/quality/causes system of Aristotle and often used the system in his descriptions of things he observed.

Albertus's greatest contribution to alchemy did not, in fact, come from Albertus Magnus. The *Book of Secrets*, which borrowed material from *De mineralibus*, was not written by Albertus but was widely attributed to him. It may have been written by a student or admirer of Albertus. At one level, it looks like *On the Properties of Things*, being a compendium of objects from the natural world, but it extended general observation to the realm of magic. The text gave details of the mythical and magical properties of gems, minerals, plants, and animals, as well as a general discussion of how to perform magic, with a strong emphasis on the necessity of trial and error and the usefulness of repeated attempts for perfecting technique.

There is in the medieval mind a kind of dual thinking about the ways of nature. Albertus Magnus and Bartholomew the Englishman were rationalists in their approach, taking Aristotle as their guide to both understanding and describing the natural world. Yet, they abandoned rationality at some point and accepted supernatural power as part of everyday life. Aristotle's work contained little of the supernatural. While there were aspects of the world that were mysterious, or even mythical, there was nothing magical in Aristotle's work, and certainly nothing that suggested that scholars could practice magic.

Part of the answer for the easy acceptance of the supernatural by the alchemists despite the rationality of Aristotle was that Christianity in the Middle Ages was full of magic. Some of this was part of the persistent pre-Christian paganism that the Church battled against but that nonetheless influenced society; it also came from a mystical heritage in Christianity. Spirits, both good and evil, were abroad in the land, along with witches, sorcerers, and mythic beasts. These ideas were reinforced by the fragmented knowledge that existed; bits and pieces of old and new knowledge filtered through the intellectual community, giving the tantalizing idea that there was both lost and hidden knowledge.

To the modern mind, these two worldviews seem to be irreconcilable, but they were not to the medieval scholar. The scholars of the medieval world who studied matter asked a basic question: How do the parts of the world work? The answer they reached was that matter existed as it did because of the active participation of God, who exerted his power both directly and through spiritual agents such as angels. Humans could and did participate in the spiritual power of the universe, just as the saints had performed miracles and witches cast spells. The guide to understanding and controlling matter was to be found in some combination of revealed knowledge (knowledge that came from God by personal revelation and the study of God's words) and the study of God's handiwork. The material world, however, was a corrupt realm, so there was also a dark side to the study of matter. The devil or evil spirits could also reveal knowledge about matter.

Alchemical compounds and elements were were also associated with other aspects of occult knowledge, such as astrology. The most basic substances were directly linked in western alchemy with celestial objects.

Celestial Object	Alchemical Material
Sun	Gold
Moon	Silver
Mercury	Mercury
Venus	Copper
Mars	Iron
Saturn	Lead
Jupiter	Tin

With this background, it is easy to understand why one of the other great alchemists of the period gained a reputation for being a magician. Roger Bacon (c.1214–1292) studied under two of the leading scholars of the medieval world, Robert Grosseteste (who worked on logic, mathematics, and Aristotelian physics) and Petrus Peregrinus (best known for his work on magnetism and practical astronomy), before joining the Franciscan order and moving to Oxford. Bacon's studies ranged from theology to technology, and he says

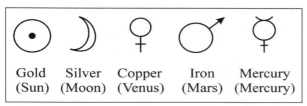

Figure 7: Alchemical symbols for materials and planets.

of his studies of the occult (meaning secret knowledge, not dark magic) that he spent 2,000 livres (a huge amount of money) on books, equipment, and supplies for experiments.

Although Bacon argued that theology was the master subject for all studies, he saw alchemy as the primary way to understand the material world. According to him, there were two branches of alchemy, speculative and practical. Speculative alchemy was the study of the generation of things from the elements and gave insight into why things such as gold, salts, and rocks were the way they were. It was a scholarly study, but Bacon proposed that the more important study was of practical alchemy,

> which teaches how to make the noble metals, and colours, and many other things better more abundantly by art than they are made in nature.[5]

Practical alchemy was based on experiment and observation, although it should not be confused with the experimental method, an idea that developed long after Bacon's time. Experiments in Bacon's alchemy were to confirm the ideas already established by speculative alchemy, not to test them or to lead to new ideas.

Bacon promoted the idea that transmutation was possible, and he used the sulfur/mercury theory of metal composition that had been developed by Al-Razi and others. While transmutation might have been important, it was not the sole focus of his work, which looked at a wide range of topics, including pharmacology and the creation of better inks. One of the most often noted aspects of Bacon's work was the idea that he independently invented gunpowder, and many textbooks and even encyclopedias repeat this story. Bacon did not actually invent or discover gunpowder, however. Rather, he knew about gunpowder from Arabic sources and produced one of the first western recipes for making gunpowder.

Bacon's work has often been portrayed as a precursor to more modern ideas about science, but in his time many of his ideas were ignored or simply not known. Bacon usually worked alone, and for a time he was kept under virtual house arrest by his religious order. In some ways, the idea of Bacon was more important in his time than the work of Bacon. He became the model of the great scholar who dedicated his life to his studies, accepting rejection and harsh conditions rather than give up his work. His interest in the practical side of alchemy and his belief that matter could be constructed was far more of a spur to later work than his actual observations, which were for the most part unavailable to other scholars.

The best known of the medieval alchemists was Paracelsus. His birth name was Philippus Aureolus Theophrastus Bombast von Hohenheim (1493–1541), and his work covered a broad range of topics, including mineralogy, iatro-chemistry (medical chemistry, covering what we would call pharmacology and aspects of physiology such as digestion), and alchemy. He was the son of a physician and spent time when he was young man in Tyrol, working in the mines and metal shops. It was probably there that he developed his idea that medical treatments should be based on a theory of metals, rather than on plants. He followed in his father's footsteps and became a physician and then settled in Strasbourg. He was called upon to treat the printer Froben, and when Froben recovered, Paracelsus's fame was spread by Froben's good friends Erasmus and Oecolampadius. He was appointed Professor of Medicine and City Physician of Basle. Paracelsus almost immediately made his presence and opinions known by burning the works of Avicenna and Galen in the town square, signifying his rejection of ancient knowledge. Only two years later, he was forced to flee Basle, having irritated many people, especially other physicians, and lost a legal battle over a fee. He then spent years wandering across Europe. He lectured and practiced when he could, but he was often hungry and destitute. He finally found refuge under the patronage of the Archbishop Duke Ernst of Bavaria, who was himself an alchemist.

Paracelsus seems to have believed in the four Aristotelian elements, but he rejected almost everything else of the ancient system. Instead, he promoted the concept of the "three principles," namely sulfur, mercury, and salt. These three principles were not exactly the materials they were named after, since they really represented concepts: sulfur was combustion and represented the soul, mercury was fluidity and represented the spirit, and salt was inertness and represented the body. By their combination and balance, they could explain the behavior of the world and, in particular, the body. Paracelsus wrote extensively, but most of his work was not published until after his death. His writing style was complex and often convoluted, full of obscure references, metaphors, and religious symbolism. For example, when describing the relation of sulfur, mercury, and salt, he said, "The Sulphur resolves itself by the spirit of Salt in the liquor of Mercury, which of itself is a liquid distributed from heaven to earth, and is the albumen of the heaven, and the mid space."[6] It would not be clear to a non-Paracelsian what this description meant.

A careful reading of Paracelsus reveals many practical aspects to his work, such as instructions for distillation and reduction, but other parts are either deliberately obscured to hide secrets or based on theoretical ideas about matter that no longer have any equivalent today. Despite the difficulty of his life and the problems he had getting his work published, he spread his ideas widely, particularly to younger students who were open to challenging the older style of medicine based on Galenic ideas.

In terms of the theory of elements, Paracelsus's contribution was not his work, although he did write about how the three principles made up most of terrestrial matter, but his role in the redirection of alchemy. He de-emphasized the concept of transmutation (although it is likely he believed it possible) and focused on practical aspects of the study of matter, particularly iatrochemistry. He also encouraged the investigation of materials through experiments. While this should not be confused with modern experimentalism, since Paracelsus included spiritualism and occult theory in his system of investigation, it was far more systematic than most alchemy tended to be. He also based his work on a conception of pure compounds, and that concept, in turn, led to work on purification and qualitative control of chemical research and production.

Paracelsus was deeply interested in experiments, another factor that fit with his willingness to challenge the old, largely Aristotelian approach to the study of nature. Aristotle had been opposed to testing nature, since he believed that the results of such tests would tell the observer only about the test, rather than about the natural world. Yet, for the alchemist, the process of mixing different substances or subjecting chemicals to various processes and observing the result was the foundation of study. As such, the concept of experimentation was promoted by alchemists, and "experimentalism" or "the experimental method" would have a profound effect on the development of science as a whole. As powerful as the experimental method later became, it would have been difficult for a student to understand the principles of experiments from reading Paracelsus's work, which was full of inference and often short on direct explanation.

A direct challenge to the old principles of natural philosophy was presented by Sir Francis Bacon, who championed experiments as the most reliable path to knowledge of nature. Bacon, like many intellectuals of his time, was interested in alchemy, and he wrote about transmutation a number of times. One of his most notable commentaries on transmutation was published the year after he died. The *Sylva Sylvarum, or a Naturall Historie in ten Centuries* (1627) was part of Bacon's great unfinished project to identify and classify all of human knowledge. In the section of the *Sylva Sylvarum* entitled "Experiment Solitary, touching the Making of Gold," he says that transmutation is possible but argues against magical or supernatural explanations for the changing of base metals into gold.

Bacon's greatest impact on science was not his alchemical work but rather his philosophical examination of epistemology, or our different ways of knowing things. Bacon was trained as a lawyer and held high office in the government of James I of England, serving as a diplomat for England, then as Attorney-General; in 1618, he became Lord High Chancellor, the highest legal authority in the land. Bacon's experience with the search for truth in legal matters, particularly in the presentation of cases at court, clearly influenced his ideas about uncovering the secrets of nature. As Lord High Chancellor, he was the person who authorized the use of torture in an era when torture was thought to provide reliable information in legal matters.

Bacon's challenge to the old system appeared in his *Novum Organum* (1620), and he wrote about his new system in *The New Atlantis* (1627). Both titles were references to Greek philosophy, since Aristotle had written a text called the *Organon*, and Plato had set his ideal society on the fictional island of Atlantis in his work *The Republic*. Bacon argued that the ancient Greeks and medieval philosophers had been unable to present a true picture of nature, since they were influenced by things like obedience to tradition, confusion over language, and anthropomorphism (seeing human characteristics in nature).

Bacon championed the "new science," which he presented as being based on induction and experimentation. Bacon concluded that reliable knowledge about nature could come only from testing nature, sometimes through destructive testing. He favored inductive reasoning, which required collecting many facts about something and assembling them before reaching a general

Figure 8: The Alchemist by David Teniers the Younger, c. 1645.

conclusion or proclaiming a scientific theory of the phenomenon being examined. The testers should be as free of bias as possible, and for this Bacon suggested a kind of division of research labor that separated the experimenters from the people who formulated principles.

Bacon's support of experiment was a direct challenge to the Aristotelian system of natural philosophy, but Bacon's idea that the results of tests and the tests themselves must be open to examination also challenged the traditional secrecy of alchemists.

Bacon's idea about the utility of experiments would have a great influence on later thinkers, such as Robert Boyle and Isaac Newton, and the term "Baconianism" is sometimes used to refer to a belief in experiments as a way of finding out the truth about nature. Although Bacon's influence was great, the specifics of his experimental system were too complex to actually use, requiring dozens of workers (many of whom did not actual know what they were working on) and the compiling of elaborate tables of information. His own experience with experiments was fairly limited, although his work on an experiment may have contributed to his death. As the story of Bacon's death is traditionally told, one day, while riding in his carriage, he had the idea that cold would reduce the speed of putrefaction in meat. He purchased a fowl, leapt out of his carriage to collect snow to stuff inside the bird, caught a cold, and died of pneumonia.

The public face of alchemy in Bacon's period is best seen in the work of David Teniers the Younger (1610–1690), a famous Flemish painter. He painted several pictures of himself as an alchemist, although whether he did alchemical work is not clear.

The paintings of alchemists are interesting in part because they show actual equipment. The painting also can be read as a kind of puzzle, containing allegorical and hermetic (meaning occult alchemical) symbols. For example, "memento mori" or reminders of death, such as skulls, frequently appear in Renaissance painting, but for the alchemist they also mean the cycle of life or the cycle of transformation. A dog represents fidelity; a candle means enlightenment or the gaining of knowledge. Although the search for secret symbols in painting can go too far—many objects end up in paintings simply because the painter likes how they look—many of the most important painters of the period were well acquainted with popular ideas in natural philosophy. This developed because perspective was part of mathematics, and the observation of nature was considered to be part of natural philosophy.

Although Bacon was interested in experiments and alchemy, he was no supporter of Paracelsus, whom he referred to as the adopted son of a family of asses (meaning the alchemical community, not his actual family) and a pretend magician. Another opponent of Paracelsus who was Bacon's contemporary was Andreas Libavius (1540–1616), who published *Alchymia* (1597). *Alchymia* has been called the first chemistry textbook, as Libavius attempted to define alchemy and then provide theory and demonstrations for chemical activity. He divided alchemy into "encheria," or operations (roughly ways to process chemicals), and "chymia," or chemical products from combinations. His work, although not completely free of mystical aspects, attacked the overt mysticism of the Paracelsians and argued for a material approach to the study of matter. Oswald Croll, a strong supporter of Paracelsus, produced *Basilica Chymia* (1609), which used the story of Genesis as a metaphor for chemical behavior. The great physician Franciscus Sylvius (1614–1672), a strong Paracelsian, established the first chemical laboratory at a university when he was Professor of Medicine at Leyden from 1658 to 1672.

One of the last and greatest of the Paracelsian disciples was Jan Baptist van Helmont (1579–1644). He made extensive use of balances to weigh materials in his experiments, opening the way to a far more quantitative approach to the study of matter. Although he was a Paracelsian, his work did not further Paracelsus's system since van Helmont, among other things, rejected both the ancient four-element system and Paracelsus's three principles. Instead, van Helmont proposed a two-element system based on air and water. He reached this position in part from an experiment he conducted with a potted willow tree. He planted a 5-pound willow in 200 pounds of earth, watered it daily for five years, and then weighed the willow and the earth. The willow had grown to weigh 169 pounds, and the earth had reduced by a few ounces. Van Helmont reasoned that since the only thing added to the willow was water and the willow had grown, vegetable matter could be composed only of water, not water and earth element. Therefore, water was a primary element, while earth could not be. This was further supported by van Helmont's belief that earthy material could be created by burning wood. It seemed to follow with perfect logic that if wood came from water, any ash or residue left by burning the wood had to come from water as well.

In another experiment, van Helmont burned 62 pounds of charcoal and found that the resulting ash weighed only 1 pound. He reasoned that the rest of the material had dispersed into the air. He called the dispersed material "geist sylvestre" ("sylvestre" meaning wood and "geist" from the Greek for chaos) and recognized that it was the same material as that produced by burning alcohol or fermenting beer and by the action of acids on shell and limestone. We get our term "gas" from "geist." In addition to "geist sylvestre," which would not burn, he identified "geist pinque," which was flammable and produced by putrefaction, or the rotting of plant and animal matter.

Van Helmont was at the cusp between alchemy and chemistry. His use of experiment and quantitative methods was innovative, and his interest in gases presaged the importance of pneumatic chemistry, which would be vital to both the practice of chemistry and the modern concept of the elements. Yet, his work was still based on a belief in transmutation and used a form of elemental theory that would have been comprehensible to Aristotle, even if he would have argued against it. Charlatans still fooled the gullible with dreams of alchemical wealth, but alchemy was seen by the intellectual elite, such as physicians and university professors, as an increasingly acceptable part of natural philosophy. In practical terms, the work of the alchemists had produced useful products such as acids and alkalis, an increased understanding of assaying in mineralogy, and an improved insight into biological processes such as digestion and fermentation.

When Nicolas Lémery (1645–1715) wrote his *Cours de Chimie* (1675), the scholarly study of matter had become firmly established. Lémery rejected transmutation and most of the spiritual aspects of alchemy, concentrating on practical aspects of chemical work. He was an atomist and experimenter and, as such, part of a new generation of researchers. Although there would continue to be alchemists who practiced in secret and sought for the elusive Philosopher's Stone well into the eighteenth century, the public face of alchemy was changing radically in the seventeenth century. It was becoming more systematic, more standardized, and more interested in the products of chemical reaction than in transmutation. There continued to be a strong and growing interest in iatrochemistry, but increasingly there was a separate discipline of research into chemistry.

The systematic study of matter using alchemical methods reached its peak in the seventeenth and early eighteenth centuries. One of the most influential alchemists of this period was Sir Isaac Newton (1642–1727). Best known for his work in physics, mathematics, and optics, he devoted more time to alchemy than to physics and believed that transmutation was a real possibility. Largely because alchemy was later shown to be impossible, this aspect of Newton's life has often been skipped over in favor of his other contributions to science. Yet, Newton's belief in alchemy was part of his effort to understand all of nature. His belief in corpuscles and his introduction of the concept of mass were part of his alchemy, not just his physics.

Despite the intellectual power of the late alchemists, transmutation could not be achieved, and it began to seem like either an impossible objective or a failed philosophy. Either way, alchemy began to lose adherents. Those who hoped to recast the study of matter in new philosophical terms rejected the idea of supernatural action in nature and focused on an understanding of chemical activity. There was a growing interest in experimentation based on quantitative knowledge rather than on qualitative knowledge, and the weighing and measuring of the elements took over from the attempts to change the characteristics of the elements.

Alchemy was also dealt a more direct blow by the events surrounding the life and sudden death of James Price (1752–1783). Price was a physician and a Fellow of the Royal Society, the oldest scientific society in the world and the most prestigious science organization in the English-speaking world. In 1782, he announced that he had transmuted mercury into silver and gold. An ingot of Price's alchemical was gold tested and shown to be real. It was even shown to the king. Price published a pamphlet on his work, which became a best-seller, but some members of the Royal Society, particularly the chemist Joseph Black and the president, Joseph Banks, were suspicious of the claims. Under pressure from both his supporters and his detractors, Price finally agreed to repeat his work for a team of observers. It was all for nothing. When the observers from the Royal Society arrived, Price swallowed Prussic acid (hydrocyanic acid) and dropped dead at the feet of the observers.

After Price's death, the Royal Society refused to investigate any other claims of alchemical transmutation. All the other major scientific societies followed the lead of the Royal Society, rejecting alchemy as a study worthy of scientific attention. In a sense, Price had put a new spin on the alchemy con game. Rather than getting people to give him money to do his work, he got people to buy a worthless pamphlet.

Although alchemists pursued a complex study that included spiritual and occult aspects, as often sought personal enlightenment rather than material control, at its heart, alchemy was always about finding that which was hidden by the confusion and corruption of the material world. In terms of both processes and products, alchemists tried to seek the elemental by systematic purification of complex materials. They did succeed in finding purer substances and in working out methods to identify materials by chemical assay. Alchemists, particularly those who followed Paracelsus, also challenged the authority of ancient knowledge, arguing that researchers could equal and surpass the abilities of the Greek philosophers. This helped open the way for new approaches to the study of nature. From the early translators of Arabic sources to the age of Newton, alchemy was one of the most important ways that people studied the physical world. Although transmutation proved to be a chimera, the by-products of alchemy included a huge range of new materials, the basic tools of the laboratory, and insights into the behavior of chemicals. In a sense, the failure to find the "prime element" opened the door to finding all the elements.

NOTES

1. While Galen's work helped Greek philosophy to survive, Galen may also have retarded the development of medicine in Europe. Cultural restrictions on the dissection of human cadavers meant that some of his observations were extrapolated from animals such as pigs and monkeys. Galen's errors, such as his belief that there were passages in the septum of the heart, were accepted as true until the sixteenth century.

2. Bartholomæus Anglicus, *De Proprietatibus Rerum* (c.1230), in *On the Properties of Things*, trans. John Trevisa (Oxford: Clarendon Press, 1975[1389]), 129.

3. Ibid., 441.

4. Ibid., 829.

5. Roger Bacon, *Opus Tertiam*, in E. J. Holmyard, *Alchemy* (New York: Dover, 1990), 120.

6. Paracelsus, *The Hermetic and Alchemical Writings of Aureolus Philippus Theophrastus Bombast, of Hohenheim, called Paracelsus the Great*, trans. A. E. Waite (London: J. Eliiot, 1894), 257.

5

CORPUSCLES AND ATOMS

As alchemy gave way to the new study of matter called chemistry, the objective of study changed. Rather than looking for the secret principles of transmutation (whether in the form of the Philosopher's Stone or the elixir of life), chemists wanted to understand the unifying principles that would explain the way matter worked. At the heart of this was a long argument, lasting more than 200 years, about the basic structure of matter. Different researchers presented ideas about the basic structure of matter, and those ideas were tested, rejected, reformulated, replaced, and tested in new ways. The ideas that survived the rigorous process of examination formed a new, open, and increasingly reliable description of matter. Key to this history were the elements. Over time, a growing group of chemists began to believe that elements were made of particles, or corpuscles. Work to identify and clarify the behavior of these particles was one of the most important steps toward a clear understanding of the elements. People who viewed matter as constructed of minute but distinct particles were called corpuscularians.

Chemistry was not pursued only for the value of pure knowledge. Solving the secrets of matter might lead to significant financial rewards. By the end of the seventeenth century, the importance of chemical production had risen to the point where it was vital to the state. Particularly in Europe, the manufacturing of gunpowder, dyes, and acids and the making of alcohol no longer relied only on local suppliers but was part of a growing international trade. Combined with the importance of iatrochemistry, understanding the material world was extending well beyond natural philosophy. It is useful to remember that although the story of the elements is mainly one of detailed research, the utility of good chemistry to industry and the power of the state was also important.

Over the next 100 years, the study of matter was also affected by the changing direction of natural philosophy. Old ideas, particularly from Aristotle, were rejected, and a new style of research began to develop, one that favored experimentation and careful measurement, rather than an examination of the qualities of matter. In more general terms, there was a move away from qualitative explanations of nature (asking why nature works the way it does) and a move toward a quantitative analysis (asking how nature works and how to measure it). Experimentation and measurement would become the hallmarks of modern science.

One of the first people to start the shift from alchemy to chemistry, as noted in the previous chapter, was Jan Baptist van Helmont. His use of experiment, measurement (particularly of weight), and his willingness to challenge existing theories made him far more like modern chemists than ancient philosophers. Van Helmont's work on gases helped to set the stage for work that would look more closely at the composition of air, and his discovery of geist sylvestre and geist pinque challenged the idea that air was a single entity. Although van Helmont's rejection of ancient matter theory and his use of measurement seem to place him at least on the way to modern chemistry, he also believed in transmutation and spontaneous generation and thus did not completely reject the esoteric or supernatural aspects of matter theory from his day.

To understand the shift in chemistry that made the modern concept of the elements possible, it is important to note that changes in natural philosophy were happening and that the study of matter was drawn along with those changes. Although the changes were widespread, two figures stand out as setting a new course for the understanding of the natural world. Francis Bacon had used the concept of induction and experiment to articulated a challenge to the Aristotelian and medieval philosophies that governed much of the examination of nature prior to the seventeenth century. An even more powerful reformation of natural philosophy came from René Descartes (1596–1650).

René Descartes' importance to science is hard to overstate. He reformed philosophy, challenging all prior philosophical systems, contributed to the development of modern mathematics by creating analytical geometry, and theorized about both the structure and composition of the universe. His most famous philosophic principle was "cogito, ergo sum," which translates to "I think, therefore I am." In other words, he broke down all knowledge to a single basic idea: that the only thing that could be known that could not be shown to depend on some more basic form of knowledge was self-awareness. It might seem as if there was a great intellectual distance between "I think, therefore I am" and the study of chemistry, but in fact the philosophical foundation of science, especially chemistry, was the drive to identify the most elemental things, those things that could not be broke down into simpler forms.

His work on physics included an explanation of inertia and the explication of many principles of optics. He took a mechanistic approach to physics, arguing that the fundamental properties were motion, mass, extension (roughly meaning space or distance), and time. Matter, according to Descartes, were those sensible objects that had the property of extension or what we might better understand as having length, width, and depth and thus occupying a measurable volume of space. Inherent in his system was a form of particle theory of existence, since he argued that the entire physical universe came to be the way it was when God introduced motion to the universe by moving the original particles.

In Descartes' system, there were three forms of matter: coarse or terrestrial matter, celestial matter, and fine matter. His coarse matter was composed of hard particles that combined to create the objects of the world. Although it is not clear whether coarse matter could be infinitely divided (there being no mathematical limit to how small length, width, and depth could be), Descartes' mechanical system of physics requires that coarse act uniformly, and coarse matter could not naturally change into celestial or fine matter.

Despite the name "celestial," this form of matter existed everywhere, not just outside the sphere of the Earth. Celestial matter was malleable and filled in the spaces around the hard particles. Fine matter, which Descartes associated with light, was even more subtle and malleable than celestial matter and filled in all remaining space. This prevented any portion of the universe from being empty space, or a vacuum.

A vacuum was a logical impossibility for Descartes, since it would be a region that could not be measured (lacking length, width, or depth), and it would also be impossible to have an immeasurable zone within a measurable zone; just as there could not be a vacuum between the planets, there could not be empty space between particles of coarse matter. Motion was communicated throughout the universe, and light (which clearly moved from source to eye) had a medium of propagation. An analogy of this system is to imagine a glass jar filled with marbles. The marbles represent the coarse matter. Adding sand (celestial matter) fills in the spaces around the marbles, but there might still be space between the sand particles. If water (fine matter) is added to the jar of sand and marbles, it will fill in the remaining space.

Descartes' particles bore a distinct resemblance to the older idea of earth, air, and fire elements. To understand this, we must remember that for Descartes, light was more primary than coarse matter, so the system had to account for the characteristics of light first and terrestrial matter second. Although Descartes was not an atomist in the strict sense of believing in indivisible particles (there was for Descartes no ultimate particle), his system treated particles as fixed entities that acted in a uniform manner that could be mathematically described. He argued that the different types matter such as gold or water were built up of particles with specific shapes in strictly uniform geometries and proportions.

Descartes' analysis of physics and matter was important in changing the direction of the study of matter. There was no place in the new physics for supernatural activity, and all aspects of the material universe were open to inspection and mathematical description. Thus, his work helped set the stage for the quantitative research of later thinkers.

Although Descartes was not an atomist, one of his contemporaries, Pierre Gassendi, was a strong atomist who revived the atomic theory of Epicurus. Gassendi wrote a biography of Epicurus in 1647, and in 1649 he published

Philosophiae Epicurus Syntagma, which outlined his Christianized version of Epicurean atomism. Gassendi argued that infinite divisibility of matter was impossible and that at creation God had "populated" the universe with atoms—particles of prime matter that were complete, perfect, and indivisible. In part on the basis of experiments he had done with mercury barometers, Gassendi also argued that there could be empty space and that the idea of matter moving through a void in fact made more sense than that of matter moving through a universe completely filled with matter. By removing the problem of what filled in the space between the atoms, Gassendi was free to picture matter in a more structural way. The matter of the terrestrial world came about by the combination of prime matter having different sizes, weights, and shapes. These different particles were held together mechanically (using a kind of hook-and-eye system), rather than by some process of attraction.

Even if Descartes and Gassendi differed on the nature of matter, they shared a view of the universe in which God created a mechanical system that did not require God's constant intervention to operate. Despite being a Roman Catholic priest, Gassendi took a materialistic view of the universe that set him at odds with many other scholars, as did his maxim "There is nothing in the intellect which has not been in the senses."

In 1654, the Oxford-trained writer Walter Charleton published *Epicuro-Gassendo-Charletoniana,* which tidied up some of the problems in Gassendi's work and brought his concept of atomism to England. Both Robert Boyle and Isaac Newton read Charleton's work on Gassendi before they did their own work on matter.

For the study of matter, Robert Boyle (1627–1691) was a pivotal thinker who introduced new methods, new tools, and a new philosophy to chemistry. Boyle was the seventh son of the Earl of Cork, and he went to school at Eton, traveled widely in Europe, and settled in Oxford, where he joined with other men interested in natural philosophy to study the important questions of the day. He would eventually be one of the founding members of the Royal Society of London and hoped to bring together the best and brightest minds in natural philosophy.

Boyle adopted the general philosophical approach to the study of nature espoused by Francis Bacon. People who followed Bacon's ideas about the study of nature tended to regard experiments as the most reliable route to knowledge and tried to be as objective about nature as possible. They also moved away from asking teleological question such as what the end purpose of matter might be. Instead, they tried to confine their questions to how nature functioned.

Boyle's research assistant was Robert Hooke (1635–1703), who would become a famous scientist in his own right. Together they performed a series of experiments that would transform not just the study of matter but the practice of science. Boyle was particularly interested in air, and his great tool for the study of air was the air pump. He looked at the relationship of pressure to volume, finding, for example, that water would boil at a lower temperature at low pressure. His experiments were compiled and published in *New Experiments Physico-Mechanicall Touching on the Spring of Air and Its Effects* (1660). It was from his work on the "spring," or compressibility, of air that the relationship known as Boyle's law originates. Using a J-shaped tube sealed at the long end and containing mercury, Boyle and Hooke demonstrated that the trapped air could be compressed or extended depending on the pressure of air on the open end. They found that pressure was inversely proportional to volume. Although Boyle did not claim that this property was universal, the principle has come over time to be a key concept in chemistry and bears his name.

The air pump also seemed to suggest that there could be regions of space that had nothing in them. The idea of a vacuum, or a void in nature, had been rejected by most natural philosophers since the time of Aristotle, but if a void could exist, it made the corpuscularian position more feasible.

Another part of Boyle's investigations with the air pump concerned the necessity of air for life. Mice placed in the air pump would pass out and eventually die if they had no air. While this was not a surprise (people understood the principle from observations of drowning), what was intriguing was that combustion and respiration seemed to be related, since fire could not survive without air. Eventually three aspects of combustion would be investigated as related actions in nature: burning, respiration, and calx formation. We would call calx formation oxidation or, more informally, rusting.

In 1661, Boyle published *The Sceptical Chymist,* setting out his challenge to the old four-element theory of Aristotle and the three-element theory of Paracelsus and opposing alchemy in general. He demonstrated that the things called elements by Aristotle and Paracelsus were not actually elemental. According to Boyle, an element had to be "primitive and simple" and "perfectly unmingled." By that, he meant that elements were the most basic (and therefore irreducible) form of matter and could not be a mixture of different simple bodies. Experimental evidence seemed to suggest that almost everything that had previously been labeled an element could be broken down into simpler parts.

Although it is clear that Boyle did not regard the traditional elements such as air and earth to be true elements, it is not clear whether he believed that any true elements existed in nature. His universe was based on a mechanical philosophy that comprised two factors: matter and motion. The corpuscles of matter had size, shape, and motion, and they were "elemental," in the sense that they were the irreducible physical particles that made up the matter of the universe. These particles could be combined in specific forms, which he called "minima naturalia." The closest equivalent we have to the minima naturalia is the modern concept of a molecule, a stable configuration of atoms. Boyle argued that since only specific forms could be constructed from the corpuscles, these minima naturalia were the most basic *chemical* substances possible, and in that sense they could be "perfectly unmingled" and were not actual elements.

Like Descartes and Gassendi, Boyle held a view of the universe that was based on a mechanical philosophy that pictures all of creation as a kind of machine operated by physical laws. Boyle succeeded in bringing this philosophy to the laboratory in a way that had not been previously attempted. Rather than simply looking at the big picture and arguing that the universe was controlled by physical rules, he demonstrated through experiment that those rules could be demonstrated at the lab bench. Boyle's work reformed chemistry by establishing mechanical philosophy as the best way to understand matter, but, in a larger sense, Boyle was also attempting to reform the study of nature. For this, historians of science remember him as a key figure in the establishment of both scientific practice and what it meant to be a scientist.

For Boyle, a scientist had the characteristics of an English gentleman: courteous, truthful, stalwart, persistent, and reliable. Like a gentleman's, a scientist's word was his bond, but, also like a gentleman, a scientist was in a privileged class and should recognize class distinctions. For example, a gentleman would not keep secrets from another gentleman, but there was no obligation to be so open with the lower classes. Thus, scientists had an obligation to share their work with other scientists, but no obligation to take nonscientists into their confidence.

When Boyle helped found the Royal Society of London around 1660 (it received a royal charter in 1662), the Society's goal was to promote the "new science" based on experimentalism and to bring together like-minded gentlemen. The Royal Society, which is still one of the most important scientific societies, helped transform the place of science in society, but it was run on a very exclusive basis. To join, a candidate had to be sponsored by someone who was already a Fellow, and all the members voted on whether the candidate was acceptable. In addition to holding meetings, the Society also published the first scientific journal, the *Philosophical Transactions of the Royal Society*. While the Royal Society created a powerful forum for scientific exchange and concentrated scientific talent in England, it had limits. John Harrison, who solved one of the most important problems in the history of navigation, finding longitude at sea, never became a member, and women were not allowed to join until 1945.

One of the greatest members of the Royal Society was Isaac Newton. Newton was elected a Fellow of the Society in 1702 and became its president in 1703, a post he held until his death, in 1727. Newton had already established his power as a scientist with his work on physics and mathematics, demonstrated principally in his famous book *Philosophiae naturalis principia mathematica*, or, more commonly, the *Principia*. Like Boyle, Newton claimed to be following Baconian method and looked at the universe from a mechanical point of view. Unlike Boyle, Newton was much closer to Gassendi on the nature of matter. His analysis of physics was based in large part on the properties of matter, particularly the property of gravity, which he argued was inherent in anything that contained mass. For many years, Newton's ideas about physics were widely known, but Newton was also very interested in alchemy and believed in transmutation. In part because alchemy was discredited later, this part of Newton's scientific work was not often mentioned by historians, but it is now clear that his alchemical work influenced both his approach to science and his belief in certain properties of matter that were used in his physics.

Isaac Newton attacked Descartes' model of the universe, but he also based his conception of matter on tiny particles, endowing each particle with mass and gravity. Newton, unlike Descartes, believed that particles were indivisible, aligning himself with Gassendi as an actual atomist. Although Newton used particles in his physics, it is not clear whether he accepted the existence of elements. His alchemical work suggests a belief in prime matter, and this matched his mechanical system, since mass and gravity were not differentiated by any chemical distinction. In other words, water and gold had different chemical properties, but the corpuscles from which they were composed had the same properties of mass and gravity, though in differing quantities. It was only in combination that outward characteristics appeared.

The particle nature of matter included light, which Newton believed was made of the finest particles. In Newton's *Opticks* (1704), he argued that all optical activity such as refraction, reflection, and transmission were based on the physics of moving corpuscles. Reflection was the rebounding of particles, while refraction was the bending of a particle's path by the influence of attraction of the matter through which the particle passed. The spectrum, or colors of the rainbow, could then be explained in terms of the separation of corpuscles of similar size and mass (all red light was one size and mass, all green another) from white light, which was a mixture of colored particles. When white light passed through a prism, the degree of attraction of the various colors of light to the glass caused them to bend slightly differently, producing the spectrum of color.

The *Opticks* was one of the landmarks of modern science. It was written in English, rather than in scholarly Latin, and it was presented as a series of experiments that the readers could do for themselves. Newton was saying, in effect, "Don't believe me just because I'm an authority. You can do the experiments and convince yourself that I am right." Newton set out a series of propositions and experiments, but he also included a series of questions and observations that he thought were interesting but needed more research. Such was the power of Newton that these questions helped shape the direction of scientific research for several generations.

The intersection of Newton's corpuscularian, alchemical, and physical interest can be seen in Book III, Part I, Question 30, where he argued that light and "gross bodies" (solid matter) must be made of particles and, further, that the particles must be transmutable. He noted that heating any gross body could make it emit light. Then he pointed out that eggs became animals and mercury be made solid and then liquid again (in a series of alchemical processes like distillation), so it was not unreasonable to assume that other forms of transmutation existed. Since he had already demonstrated the particle nature of light, it logically followed that the production of light from gross bodies meant that they were ultimately transmutable; as Newton put it, "why may not Nature change Bodies into Light, and Light into Bodies?"[1]

One of the consequences of Newton's corpuscular theory was the idea that sight was caused by a particular type of sense of touch. Vision, according to Newton, occurred when light particles struck objects and were reflected into the eye, exerting pressure on the optic nerve. Newton, perhaps taking his science to a bit of an extreme, attempted to prove his theory of vision by slipping a dull knife behind his eye to press on the nerve. This, he reported, resulted in the appearance of bright flashes of light, sparkling stars, and dancing flames, thus proving his argument.

By the time of Newton's death, in 1727, the corpuscular theory of matter (despite the variations) had come to dominate the study of nature. What was less clear was whether elements played any role in understanding the natural world or whether it was better, following the various mechanical philosophies of Descartes, Gassendi, Boyle, and Newton, to consider the world in terms of motion and mass. In a sense, the rise of physics reduced the number of elements from four or three to one. The problem for the natural philosophers was that they could not quite agree what characteristics that one element could have that would result in the huge range of matter in the real world. In terms of science, the triumph of experiment and the "new science" opened the door to a far more open and systematic examination of matter.

PNEUMATIC CHEMISTRY

One of the big riddles for the scientists of the seventeenth and eighteen centuries was the process of combustion. In addition to burning, combustion seemed to be related to other things, like respiration and rusting. Understanding "airs," or what we would call gases, would reveal important aspects of the function of nature. It would also have a profound effect on matter theory.

Studying airs represented a complex experimental challenge, since airs were usually invisible, difficult to produce and collect, and hard to analyze. Collectively, the study of gases was called "pneumatic chemistry," from the Greek *pneuma,* meaning breath. The existence of air as an element was fundamental to the Aristotelian system, but the existence of particular types of airs also went back to the Greeks and earlier. The different types of airs included things like swamp gas, smoke, and fumes produced by the action of various chemicals, such as that of acids on iron. These were thought to be corruptions of the elemental air, which was colorless, odorless, and almost, but not quite, weightless.

Robert Boyle and his assistant Robert Hooke created a way to evacuate a glass dome so that they could do experiments concerning air, treating air as a collection of very fine particles that could be compressed or expanded to fill volume, a characteristic that matter in other forms (liquid and solid) could not duplicate. They also used the air pump to study combustion, noting that burning stopped when air was removed. This strongly suggested that air was needed for combustion, but it could not be the whole answer, since they also noted that gunpowder could burn under water. In addition, Boyle observed that metals gain weight when heated in air. These kinds of results led Boyle and other scientists to suggest that there was a substance of combustion. Boyle identified "igneous particles," while Hooke called them the "nitro-aerial spirit."

In 1667, Johann Becher (1635–1681), a chemist interested in the origins of metals, published *Physicae subterraneae.* He argued that there were three "earths": *terra fluida* (mercurious earth), *terra lapidea* (vitreous earth), and *terra pinguis* (fatty earth). It was terra pinguis that produced combustible properties. Anything that burned must contain some proportion of terra pinguis. Although these three earths could be deduced, they did not exist independently in nature. In specific combinations, they were held together by affinity (a natural attraction in fixed proportions) to produce the most stable substances, such as gold or silver. These substances could not be broken down and were the simplest material of chemistry.

Georg Stahl (1660–1734) based his own theory of matter and combustion on Becher's work but changed the name "terra pinguis" to "phlogiston" (from the Greek word *phlogos,* for flame) and extended the principle of combustion to include calcination and metal formation, or what are called today oxidation and

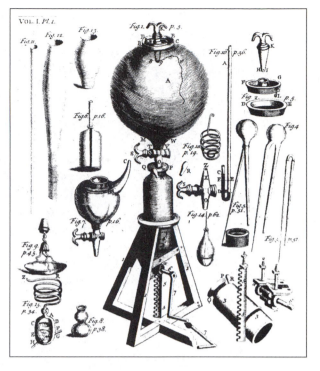

Figure 9: Robert Boyle's air pump.

reduction. For some chemists, phlogiston was an element, somewhat like fire in the Aristotelian system, but for most researchers (including Stahl) phlogiston was a separate class of matter. One of the reasons that phlogiston was considered different from regular matter was that it appeared to be mixed with other things and caused no change until released in combustion. For example, in Stahl's system, you could have a reversible transformation of metals.

Metal $\rightarrow$ Calx + phlogiston
(iron containing phlogiston)   (rust)(phlogiston released)
Calx + phlogiston $\rightarrow$ Metal
(rust)(phlogiston added)   (iron containing phlogiston)

The phlogiston theory was a powerful tool to explain not only combustion but also respiration, rusting, and the properties of acids and bases. It could also be used to explain different rates of combustion and the end of combustion, on the basis of the quantity of phlogiston present and the degree to which the surrounding material could absorb phlogiston.

Theories of combustion, particularly the widely held phlogiston theory, focused work on pneumatic chemistry. Stephen Hales (1677–1761), an English biologist, introduced an important tool for the study of gases called the pneumatic trough. He used a bent gun barrel sealed at one end to collect the gases released by the combustion of dozens of different substances, including blood, tallow, and sugar. He would place the material in the closed end of the tube and place that end of the tube in a fire. The open end went under a bottle filled with water and inverted over a bucket of water. As the gas was produced, it pushed out the water in the bottle and thus could be collected. Hales was particularly interested in "fixed air." He observed that plants absorb air through their leaves and reasoned that it must be converted into solid matter, hence "fixed" or made permanent.

Unfortunately, Hales did not look at the different properties of the gases he collected, treating each as essentially the same. The careful study of individual gases was advanced by Joseph Black (1728–1799), Henry Cavendish (1731–1810), and Joseph Priestley (1733–1804).

Joseph Black was the son of a Scottish-Irish wine merchant who settled in Bordeaux. Black was educated at Belfast and studied at Glasgow University to become a physician. Although he earned an M.D., he worked most of his life as a professor of chemistry. Black demonstrated that fixed air (or what is now called carbon dioxide) could be obtained by many different chemical reactions and existed in exhaled air from respiration. He further demonstrated that it was different from atmospheric air and also a component of the atmosphere, thereby proving that air was not a single element but a mixture. The idea that air was not an element greatly undermined the Aristotelian system of four elements. In addition to his work on fixed air, Black helped introduce the quantitative study of gases, working out ways to collect and weigh gases. He also explained the phenomenon of latent heat, or the quantity of heat needed to produce a change in state of a solid to a liquid or a liquid to a gas.

Henry Cavendish was one of the great English amateur scientists. He came from a wealthy family from the line of the Dukes of Devonshire, although he was not personally wealthy until later in his life. He went to Cambridge to study but never graduated. Cavendish was a very private person and conducted most of his work for his own interest, publishing only three major papers, one on his discovery of "inflammable air" (what we call hydrogen), one on the relation of fixed air, calcareous matter (calcium), and water, and one a study of the composition of the atmosphere.

Although combustible gases had long been known to exist, Cavendish isolated inflammable air and suggested that it was phlogiston, or the substance of fire. He also demonstrated that inflammable air was present in a wide variety of materials, such as alcohol and metals exposed to acid. This reinforced the idea he had identified the material principle of combustion. It was also in accord with his analysis that the atmosphere was composed of a number of gases and that the distribution of those gases was uniform.

Cavendish's heritage is commemorated in the great Cavendish Laboratory in Cambridge. It was William Cavendish, 7th Duke of Devonshire and Chancellor of the University, who established the laboratory, in 1874.

The champion of gas isolation was Joseph Priestley. Priestley's personal history was a tale of trial, achievement, and political conflict. He was from a family of weavers who lived near Leeds, but he went on to study for the ministry. While he worked as a minister, he was also studying the latest developments in science and was encouraged by his friend Benjamin Franklin to write a history of electricity. He went on to produce works on chemistry, politics, and theology. Priestley was a Dissenter and wrote against the Church of England. He was also sympathetic to the claims of the American colonies. In 1791, a mob burned down his house in Birmingham, and Priestley fled to Pennsylvania, where he stayed until he died.

Priestley devised ways to collect gases that was based on the use of what was called a "pneumatic trough." Priestley inverted a glass container full of water and placed it in a pan of water. Gases were then allowed to bubble up through the water and displace the water in the container. In addition to the traditional method of bubbling gases through water, Priestley also used mercury for gases that dissipated in water.

Among the gases Priestley isolated and studied were nitric oxide, nitrogen dioxide, ammonia, and hydrogen chloride. One of his most important areas of study was "dephlogisticated air" (what we would call oxygen). He generated dephlogisticated air by heating mercuric oxide, a red powder, and collecting the gas in a pneumatic trough filled with mercury, rather than water (since the gas was absorbed by water). A candle exposed to the gas burned brilliantly. Unlike other gases that supported combustion, dephlogisticated air was also suitable for respiration. Priestley reached this conclusion by studying mice placed in sealed containers with various gases. Priestley reasoned that dephlogisticated air was better for combustion because it contained so little phlogiston that it could absorb a great deal of the phlogiston being generated by respiration or combustion. Priestley recorded his major discoveries in his massive three-volume work, *Experiments and Observations on Different Kinds of Air* (1774–1790).

In a further blow to the ancient element theory, in 1783 Cavendish combined his inflammable air with Priestley's dephlogisticated air and produced water. Although by the eighteenth century few chemists believed that water was an element, it was not clear what things combined to produce water. Cavendish's experiment showed that it was two specific gases that, when combined, produced water. Water was therefore not an element, since it was composed of more than one substance, but the gases seemed to many to be simpler and thus more likely to be elemental.

The success of pneumatic chemistry opened the door for a much more detailed study of matter by allowing people to conduct experiments on the most rarified state of matter. Although the existence of different types of airs did not prove the existence of elements (the airs might be broken down by later experimenters), many

Figure 10: Joseph Priestley's pneumatic trough used for collecting gases.

scientists felt that gases were closer to the corpuscular foundation of matter. In what might seem like a paradox, by proving that a substance like water was not an element, scientists suggested strongly that elements did indeed exist. The ability to identify and quantify the mixture of gases in water made it much more conceivable that matter could be identified by fixed and unchanging conditions that were unique to a particular kind of matter.

NOTE

1. Isaac Newton, *Opticks* (New York: Dover, 1979), 375.

6

BRINGING ORDER TO CHAOS

While work on gases extended knowledge about the constituent parts of the chemical world and gave important tools to those studying matter, it created more chaos than it resolved. Instead of three or four elements of the older systems or the one prime element of the mechanical philosophers, almost everything seemed to be made up of something else, and what, exactly, that something else might be seemed harder to identify than had been believed. Water was made up of gases, the atmosphere was full of different airs, and combustion was a strange substance that could flow from one place to another or saturate a volume of space. Although the idea of alchemical transmutation was fading from serious consideration in the eighteenth century, ideas about the transformation of matter continued to be confusing. For example, it seemed that there was some relationship between respiration and combustion, but there was clearly a difference between organic and inorganic matter. While a chemist might be able to demonstrate the reversible transformation of metal into calx and back, the same was not true of organic matter. A piece of wood could not undergo combustion and then be restored from ashes to wood.

On top of these problems remained the age-old problem of communication. Although there was more open communication of ideas and discoveries, partly as a result of work by Boyle and the creation of the Royal Society, nomenclature continued to be a problem. Each chemist had his own name for things, and there was no generally accepted system for naming new things. This made it hard to know if the work of one chemist confirmed the work of another or concerned something completely different.

Into this state of affairs came Antoine Laurent Lavoisier (1743–1794). Lavoisier was the son of a wealthy Parisian lawyer and was sent to the best schools. Early in his life, he showed a strong interest in science, particularly chemistry. In 1768, he was elected an adjunct member of the Académie

Royale des Sciences and became a regular member a year later. This was an astonishing achievement for a 25-year-old, since the limited number of Académie positions were usually awarded to well-established scientists with long records. He was frequently called on by the Académie and the government to investigate issues ranging from education, prison reform, and weights and measures to ways to improve farm yields.

The Académie des Sciences was not a French version of the Royal Society (which was self-selecting and largely composed of amateur researchers) but a state-sponsored and -supported organization. It had been created in 1666 by Jean-Baptiste Colbert, who was the controller general of finance. His idea had been to harness the power of French intellectuals for the betterment of the state. With the help of royal patronage, the Académie des Sciences provided salaries for more than 100 thinkers, about 60 of whom worked in natural sciences. In exchange for the financial support, the members of the Académie des Sciences did work for the Crown and the government. In addition to paid positions, the Académie also offered prizes for various projects.

In 1775, the government appointed Lavoisier as one of four commissioners of the Gunpowder and Saltpeter Administration. As the *régisseur des poudres*, he was the person who oversaw the production of gunpowder throughout France. In addition to a government salary, he received a house and laboratory space at the Arsenal in Paris. In a few years, Lavoisier's work on the chemistry of gunpowder and production methods took French powder from being the worst in Europe to being the best.

Lavoisier's government commitments did not slow his own investigations. Because he was personally wealthy, both from family money and as a partner in the Ferme Générale (a private company that collected taxes for the government), Lavoisier had the best equipped laboratory in Europe. Everything from the most delicate balances for measuring the mass of gases to powerful furnaces was at his disposal.

With these tools, he began to investigate gases. In particular, he looked at phlogiston theory. One of the great problems of phlogiston theory was that some substances, such as wood, lost weight when burned, while metals gained weight during calcination. The old explanations of how phlogiston worked either did not acknowledge this problem or were based on the idea that phlogiston was an "imponderable fluid," meaning that it was not ordinary matter. An imponderable fluid was a substance that behaved like a liquid (it could flow and be divided) but had no weight. This did not satisfy Lavoisier. While ash residue was lighter than the wood it came from, suggesting that phlogiston had been removed, how was it possible with a metal to take something away (the phlogiston) from a substance and make it heavier than it was to begin? Although others had noted this problem earlier, researchers had not carefully measured or even explained it as a kind of buoyant property of phlogiston that acted like hot air.

In 1783, Lavoisier submitted a study of phlogiston entitled *Reflections on Phlogiston* to the Académie. By careful experiment and logical deduction, he

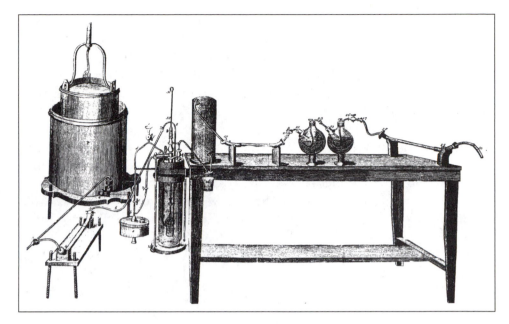

Figure 11: Lavoisier's constant pressure gas pump and reaction vessels.

destroyed the phlogiston theory by demonstrating that all the effects attributed to the mysterious substance could be accounted for by the action of oxygen, a substance that could be identified, created, measured, and used in experiments. Lavoisier replaced phlogiston theory (based on a system that required a separate type of matter) with one based on unified form of matter in which chemical reactions with a specific gas accounted for the observed events. The substance known as dephlogisticated air, which Lavoisier renamed "oxygine" (after the Greek for "acid former"), did not have mysterious properties, nor was it the spirit of combustion.

Not everyone accepted Lavoisier's new system, particularly Joseph Priestley. Defending phlogiston to the last, Priestley wrote *The Doctrine of Phlogiston Established* (1800), trying to restore phlogiston as the key to combustion, but it was already too late. Chemists, even if they were not completely convinced by Lavoisier, recognized that the old theory had too many contradictions to be considered seriously. In addition to the uncertain characteristics of phlogiston, scientists were becoming skeptical about the existence of unique. Several key chemists, including Claude Berthollet (1743–1822) and Louis-Bernard Guyton de Morveau (1737–1816), adopted Lavoisier's position and helped to spread Lavoisier's system of detailed and quantified experiments. In particular, Antoine Fourcroy's (1755–1809) texts *Leçons élémentaires d'histoire naturelle et de chimie* (1782) and *Système des connaissances chimiques* (1804), based on Lavoisier's ideas about acids, bases, salts, and elements, was used by the next generation of chemists as the foundation of modern chemistry.

What was needed to clarify chemistry, however, was more than solving the problem of chemical combination in combustion. Lavoisier, along with

Berthollet, Morveau, and Fourcroy, decided to formulate a new nomenclature for chemistry based on Lavoisier's work. They believed that to reform chemistry, it had to be reformed completely, with a new language of chemistry for a new system of chemistry. Their work was published in 1787 as *Méthode de nomenclature chimique*. It is from this work that the modern system of naming chemicals developed. It introduced word endings like *–ic* to represent acids and *–ide* to indicate oxides.

The nomenclature book was not widely adopted, however. In a sense, it was too specific, and it assumed the correctness of Lavoisier's system without providing a clear explanation of that system. Lavoisier, eager not only to convince skeptics but to influence the next generation of chemists, went on to publish his *Traité élémentaire de chimie* (Elements of Chemistry) (1789). This book was a best-seller and had a profound influence on the development of chemistry. It was quickly translated into English, German, Dutch, Italian, and Spanish. In it, Lavoisier introduces his system, with examples and Illustrations, as well as his nomenclature. He explains the importance of nomenclature in the Preface.

> The impossibility of separating the nomenclature of a science from the science itself, is owing to this, that every branch of physical science must consist of three things; the series of facts which are the objects of the science, the ideas which represent these facts, and the words by which these ideas are expressed. Like three impressions of the same seal, the word ought to produce the idea, and the idea to be a picture of the fact. And, as ideas are preserved and communicated by means of words, it necessarily follows that we cannot improve the language of any science without at the same time improving the science itself; neither can we, on the other hand, improve a science, without improving the language or nomenclature which belongs to it. However certain the facts of any science may be, and, however just the ideas we may have formed of these facts, we can only communicate false impressions to others, while we want words by which these may be properly expressed.[1]

Lavoisier's concern with the proper names of things went beyond simply getting people to use the same name for the same substance in order to facilitate communication. He held a belief, shared by many intellectuals of the period, that the words one used shaped the thoughts one could have. Thus, his system of nomenclature contained within it his system of chemistry. In other words, if you accepted his names, you were likely to accept his way of thinking about chemistry. By dominating the production of texts on chemistry, the next generation of chemists literally grew up with the new system.

Before he could proceed to the behavior of matter, particularly his ideas about chemical combination, Lavoisier had to state the basis of his concept of elements.

> All that can be said upon the number and nature of elements is, in my opinion, confined to discussions entirely of a metaphysical nature. The subject only

furnishes us with indefinite problems, which may be solved in a thousand different ways, not one of which, in all probability, is consistent with nature. I shall therefore only add upon this subject, that if, by the term *elements*, we mean to express those simple and indivisible atoms of which matter is composed, it is extremely probable we know nothing at all about them; but, if we apply the term *elements*, or *principles of bodies*, to express our idea of the last point which analysis is capable of reaching, we must admit, as elements, all the substances into which we are capable, by any means, to reduce bodies by decomposition. Not that we are entitled to affirm, that these substances we consider as simple may not be compounded of two, or even of a greater number of principles; but, since these principles cannot be separated, or rather since we have not hitherto discovered the means of separating them, they act with regard to us as simple substances, and we ought never to suppose them compounded until experiment and observation has proved them to be so.[2]

Lavoisier was very interested in the practical aspects of chemistry, and his definition of the elements is very pragmatic. Essentially, it says that those things that have not been broken down into simpler substances should be considered elemental. Rather than taking a theoretical or philosophical stand, his definition was based on what a chemist could do in a laboratory. It also meant that if later work did break down a substance listed as an element into more basic parts, the discovery would not threaten the larger system at all.

The first section of the book deals with the concept of heat and gases. Lavoisier adopts van Helmont's term "gas" for "elastic fluids," replacing the older "airs." Although Lavoisier had discarded the imponderable fluid phlogiston, he could not completely abandon the idea that heat was a substance. For heat, he introduces "caloric" as a different but, he hopes, more rational imponderable fluid. Lavoisier and Pierre Simon Laplace (1749–1827) said that caloric could flow from a hot region to a cooler one. The addition of caloric to something caused it to expand, which would, for example, account for the observed expansion of metals. With his partner Laplace, who was a brilliant physicist, Lavoisier constructed a device to measure amounts of heat, as opposed to measuring temperature. The distinction between heat and temperature was confusing, particularly before the development of a kinetic theory of heat, but scientists of the time recognized that a boiling kettle, while very hot at 100° centigrade, did not contain as much heat altogether as a lake did, even if the temperature of the lake was only 15° centigrade. Laplace and Lavoisier's device, called the calorimeter, could quantify the amount of heat. By measuring the amount of water that could be melted from ice by some heat source (such as combustion, a chemical reaction, or even the heat generated by a mouse), a careful measurement of heat could be calculated. Although the concept of caloric was later abandoned with the introduction of the kinetic theory of heat (based on the motion of atoms), Lavoisier's idea about measuring heat continues in our use of the calorie as a measurement of heat. We still use the calorie (really a kilocalorie) as a measurement of food energy. It is based on the amount of heat produced by the complete combustion of a fixed amount of food.

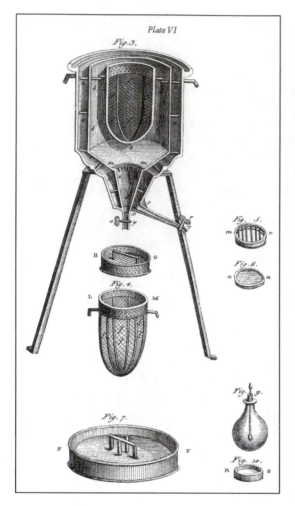

Figure 12: The calorimeter.

The second section of the *Elements* starts with what Lavoisier calls "simple substances" and provides a list of these elementary substances. Many of the 33 substances, ranging from antimony to zinc, would prove to be in their simplest form. These are materials that had not been broken down into simpler substances. He treated them as elements, although he was careful to leave open the possibility that chemicals on the list might be broken down in the future. In particular, he included several substances such as magnesia and alumina, which he suspected were oxides of unidentified metals. In addition to providing a working definition of elements and a list, he also attempted to clarify the names for the different substances, listing his new name and a variety of old names.

Another concept that was at the heart of Lavoisier's work became known as the law of conservation of mass. He assumed that in any chemical reaction, there would be no gain or loss of matter. Although others had used this idea, the general methods of chemical work were often so inexact that it was impossible to be sure what masses were involved. This was partly a matter of theory and partly a problem of equipment. If, for example, a researcher who measured the weight of ash left from a wood fire would find that the ash weighed much less than the original wood. If your theory said that the phlogiston in the wood was expelled into the atmosphere, you might not bother attempting to measure anything else, since your theory accounted for the observation. Further, if you had no equipment to enclose and control the whole combustion experiment or to measure the mass of the gases before and after combustion, it would be impossible to for you to make all the necessary measurements even if you wanted to. Fortunately, Lavoisier had the equipment to measure things like the mass of gases. He was therefore able to demonstrate that in chemistry matter is never destroyed; it can change composition (e.g., by forming oxides) or state (e.g., from liquid to gas), but all chemical reactions must balance.

In addition to all the things that Lavoisier contributed to chemistry, his work advanced the idea of analytical chemistry. Lavoisier carefully measured everything in his experiments, and his laboratory was equipped with the most precise measuring instruments available, such as balances to measure mass, barometers to measure pressure, and displacement cylinders to measure volume. This in turn helped to establish one of the most important principles is chemistry—the law of definite proportions. This law states that the proportion of elements in a compound are always the same, regardless of the process that produces the compound.

One of the most confusing aspects of chemical combination was that it seemed that there was a broad range of products from a single chemical interaction.

Lavoisier's work was interrupted by the French Revolution. Although he was at heart a reformer, he was closely associated with the royal government and was a partner in the hated tax farm system. That alone would have made life difficult for Lavoisier, but he was also directly attacked by Jean Paul Marat, one of the leaders of the Terror. In the violent times, Lavoisier was arrested, tried by a revolutionary court, and sent to the guillotine on May 8, 1794.

TABLE OF SIMPLE SUBSTANCES.

Simple substances belonging to all the kingdoms of nature, which may be considered as the elements of bodies.

New Names.	Correspondent old Names.
Light	Light.
Caloric	Heat. Principle or element of heat. Fire. Igneous fluid. Matter of fire and of heat.
Oxygen	Dephlogisticated air. Empyreal air. Vital air, or Base of vital air.
Azote	Phlogisticated air or gas. Mephitis, or its base.
Hydrogen	Inflammable air or gas, or the base of inflammable air.

Oxydable and Acidifiable simple Substances not Metallic.

New Names.	Correspondent old names.
Sulphur	The same names.
Phosphorus	
Charcoal	
Muriatic radical	Still unknown.
Fluoric radical	
Boracic radical	

Oxydable and Acidifiable simple Metallic Bodies.

New Names.		Correspondent Old Names.
Antimony		Antimony.
Arsenic		Arsenic.
Bismuth		Bismuth.
Cobalt		Cobalt.
Copper		Copper.
Gold		Gold.
Iron		Iron.
Lead	Regulus of	Lead.
Manganese		Manganese.
Mercury		Mercury.
Molybdena		Molybdena.
Nickel		Nickel.
Platina		Platina.
Silver		Silver.
Tin		Tin.
Tungstein		Tungstein.
Zinc		Zinc.

Salifiable

Figure 13: Lavoisier's table of simple substances from *Elements of Chemistry* (1790).

One of France's other great chemists, Claude Louis Berthollet (1748–1822), had argued that the composition of a compound would vary within a large range of possibilities. This was not in complete opposition to Lavoisier's results about specific combinations, but it did create a theoretical problem, since it made it possible to create a large, or even infinite, number of substances. This idea was supported by observations of things like the progressive color change of oxides during the course of a reaction.

On the other side of the argument was Joseph Louis Proust (1754–1826), who argued that only specific and fixed combinations were possible. Through carefully controlled and analytical experiments, Proust demonstrated that there were fixed proportions of matter in chemical substances that were made of a combination of simple substances. What had confused scientists was that in many reactions, more than one final product was present, such as a mixture of two types of tin oxide. Since the absolute quantity of each oxide depended on a large range of variables for any given experiment, it had seemed that there was a range of end products, rather than varying quantities of two fixed products.

Although it might seem paradoxical, unraveling the problem of chemical combination was an important step toward understanding elements. If matter was variable or came in a spectrum of possible conditions, no concept of elements was possible, since there could be no ultimate particles or corpuscles. When Proust was able to demonstrate definite proportions, it strongly suggested that the basic units of matter were unchanging and had specific properties.

The rise of analytical chemistry as promoted by Lavoisier and his supporters led to a burst of activity by chemists and other researchers. One of the foremost researchers was Martin Heinrich Klaproth (1743–1817). Klaproth started his professional career as an apprentice apothecary but rose to the position of professor of chemistry at the University of Berlin. He was an early German convert to the Lavoisier's system and believed in carefully controlled experiments. His background as an apothecary probably influenced his development of analytical chemistry, since apothecaries were constantly concerned about methods for determining the composition and purity of the materials they used. In an era when quality control of chemical production was largely nonexistent, it was up to apothecaries to determine whether the materials they dispensed were what they were supposed to be. Making mistakes could be dangerous, causing illness or even death.

Over his career, Klaproth examined thousands of substances, but he is best remembered for his work on "earths" (what are now called metallic oxides), particularly the "rare earths." These materials were called rare because they were literally rare, occurring in tiny amounts among minerals that were themselves very scarce. Today, the rare earth elements are numbers 57 to 71 on the periodic table.

In 1789, Klaproth isolated zirconia from zircon, and in the same year he discovered uranium in the ore pitchblende. He confirmed the existence of strontia (discovered by Thomas Charles Hope [1766–1844]) in 1795 and discovered titanium that same year. In 1797, he isolated chromium, while in 1789 he announced the discovery of tellurium. Klaproth shares a three-way discovery of cerium with Wilhelm Hisinger (1766–1852) and Jöns Jakob Berzelius (1779–1848).

These discoveries were based on meticulous and time-consuming laboratory work. It was like doing a crossword puzzle with some of the clues missing. Filling in the words with clues partially revealed the unknown words. Known compounds and elements had specific properties such as solubility, melting point, and chemical reactivity. By going through a series of steps that removed material with known properties, the researcher was left with the unknown material. Further tests then established that the unknown substance had unique characteristics, thus confirming that it was a new element. To do this, raw ore had to be ground to a fine powder, then washed in various solutions to remove impurities. Partial distillation was used to drive off more impurities, while heating in the presence of various chemicals that would combine with known elements further purified the sample. What remained was often an oxide, or a mix of the pure element with oxygen. Removing the oxygen was almost impossible with the equipment Klaproth and the chemists of the period could use, but because the characteristics of oxygen were so well known, characteristics such as atomic weight, specific gravity, and melting point of the pure element could be deduced.

Lavoisier and his supporters believed that elements were the foundation of chemistry and the discovery of new elements fit their system. The new chemistry was more flexible than the older matter theory of the Greeks or the alchemists. In the old systems, any demonstration that an element was not truly elemental called into question the entire system. By the end of the eighteenth century, earth, water, and air had all been shown to be compounds, largely wiping out the Aristotelian system. The extension of the elements beyond mercury, sulfur, and salt was equally devastating to the Paracelsian model. In contrast, the new chemistry left open the possibility that future work might show that oxygen was a compound, but such a discovery would not destroy the theory of chemistry. The new chemistry was based on a series of principles about the process by which knowledge was discovered and confirmed, rather than statements about the structure of the universe.

In a sense, the chemists also avoided the physical issue of whether elements were necessary in a mechanistic universe by focusing on the functional aspects of the material world. Simply by establishing that elements had unique characteristics, they could differentiate matter. Elements could be identified, classified, isolated, and used in experiments. If Descartes, Boyle, and Newton pictured matter as some form of prime particles differentiated only by proportion, mass, or size, it did not really matter to chemists. So long as the resulting products had unique and unchanging properties, the designation "element" was both practical and scientifically justified. While what made an element unique would continue to be a profound scientific question, by the beginning of the nineteenth century chemists were convinced that elements were real and that they served as a sound basis for scientific research.

NOTES

1. Antoine Lavoisier, *Elements of Chemistry* (New York: Dover, 1965), xiv.
2. Ibid., xxiv.

7

MAKING ELEMENTS ELEMENTAL

Although the work by Klaproth, Lavoisier, and the other chemists of the age expanded knowledge of the elements and transformed chemistry, understanding why elements were elemental was still a mystery. As a functional distinction, elements could not be decomposed or broken down into subsidiary components. In general, all chemists agreed that the materials generally known as elements met this definitions, but it was not clear from a structural or chemical reason why oxygen and hydrogen were elements but water was not. For many people who worked with chemicals, such as apothecaries, brewers, and dye makers, this knowledge was not particularly important. What was important was the ability to produce consistent products. For others, the clarification of the existence of elements and the much greater reliability of chemical research that it produced only heightened their desire to determine what made matter behave the way it did.

Some of the first steps toward understanding how the basic corpuscles of matter behaved came from what might seem like an unlike source: meteorology. John Dalton (1766–1844), born in Manchester, began to study weather in 1787. He also studied Newton's *Principia* and as a result was conversant with the concept of the inverse square law. Newton had used it to show that the attraction of gravity decreased as the square of the distance between objects. This mathematical relationship also applied to a range of other physical phenomena, such as luminosity. When considering questions about precipitation—why there was fog, rain, and snow—he discovered that the quantity of water vapor in a gas was independent of the type of gas but dependent on the temperature of the gas.

This led him to question the idea that the atmosphere was a chemical compound. Instead, he began to think of it as a mixture of different gases that existed independently. To understand the distinction, consider a cup of hot chocolate. The cup seems to contain a uniform substance that is made by combining chocolate with milk. We recognize that there are two original components but treat the resulting drink as a single substance. On closer examination, what really is happening in the cup is that there are particles of chocolate dispersed through the milk. They never really combine into a new substance. Similarly, atmospheric air had been treated like a single substance made up of different original components, such as oxygen, nitrogen, and carbon dioxide. Dalton wondered whether changing this concept would explain the meteorological problem he had discovered.

In 1801, Dalton formulated the law of partial pressure. He argued that the pressure exerted by one type of gas in a mixture was independent of the pressure exerted by another type of gas. In a fixed volume, the observed pressure was then the total pressure made up of the sum of pressures exerted by each gas.

While this fit well with his idea that the atmosphere was a mixture rather than a compound, it led to a new puzzle. He knew that the different gases that made up the atmosphere had different densities, so that, for example, oxygen was denser than nitrogen. Why, then, weren't there layers of gases over the earth? If you mixed water and oil, the oil would form a layer on top of the water. Dalton reasoned (in part from his knowledge of Newton) that the reason was that the particles of gas would repel particles of the same gas according to a force of repulsion that fell off at the square of the distance. One type of gas particle was unaffected by other types of gas particles. Thus, the particles of all gases in the atmosphere would disperse everywhere and not separate into layers. The difference in gases seemed to be physical, and, following this line of thought, in 1803, Dalton produced a table of the relative weights of what he called the "ultimate particles."

At about the same time that Dalton was contemplating the structure of the atmosphere, William Henry (1774–1836), a close friend, was examining the solubility of gases in liquids. He noted that the amount of a gas dissolved in a liquid was independent of pressure. When he discussed this finding with his friend, Dalton concluded that the dissolving process must then be physical, rather than a form of chemical combination. He also noted that lighter particles seemed to be the least absorbable and heavier ones more easily absorbed. This led Dalton to think more about the size and weights of particles and to look for a way to link the physical nature of the particles with their behavior. He established the relative weights of his particles by arbitrarily setting hydrogen to 1.0 and comparing other things like oxygen, ammonia, and water to hydrogen.

With these ideas as a foundation for understanding matter, Dalton went on to publish his *New System of Chemical Philosophy* (vol. 1, part 1, 1808, part 2, 1810; vol. 2, 1827). In it, Dalton presented his new theory of atomism. Dalton made four assumptions about atoms and the compounds built up from these ultimate particles.

First, he started with the idea that the simple particles were indestructible and that they remained unchanged even when in combination with other things. Although this was not easily proven as universally true in Dalton's time, it was necessary for reversibility, such as the mixing of hydrogen and oxygen to produce water and the decomposition of water back into hydrogen and oxygen. The decomposition of water had been demonstrated by a number of scientists, including Lavoisier.

Second, the particles in Dalton's system were solid and indivisible and remained unchanged regardless of state. Thus, ice, liquid water, and steam (water vapor) all comprised the same particles, but there were different relationships between the particles.

Third, Dalton assumed that there were different types of atoms. This went against the corpuscular theories of Newton and Boyle, who had thought that there was only one primary particle and that all the simple elements (although normally indivisible) were made of particular combinations of primary particles.

Dalton's fourth assumption was that atoms had specific atomic weights and could thus be distinguished from one

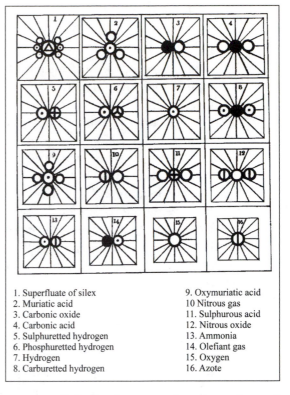

1. Superfluate of silex
2. Muriatic acid
3. Carbonic oxide
4. Carbonic acid
5. Sulphuretted hydrogen
6. Phosphuretted hydrogen
7. Hydrogen
8. Carburetted hydrogen
9. Oxymuriatic acid
10 Nitrous gas
11. Sulphurous acid
12. Nitrous oxide
13. Ammonia
14. Olefiant gas
15. Oxygen
16. Azote

Figure 14: Dalton's elements and common "atoms." From John Dalton, *New System of Chemical Philosophy* (1808).

another by a simple test. The test was simple, at least in theory: All you needed to do was collect a specific volume of an element and use a balance to measure its mass. In practice, measuring the specific atomic weight would be a major challenge for scientists. Not only was collecting and measuring some of the elements very difficult, but also the results could be expressed only as ratios, comparing the mass of one element to another. What scientists wanted was a measurement of the mass of an individual corpuscle of an element.

Together, these assumptions helped to clarify the concept of elements and chemical combination. Compounds, using Dalton's system, resulted from the joining together of atoms of different elements. A one-to-one combination was a binary combination, and Dalton represented it by placing two circles together. More complex arrangements could then be built up from the simpler ones. Dalton may have been one of the first people to build actual physical models of elements and compounds, but, since so many of Dalton's original notes and records were destroyed during World War II, we know very little about how he used his models.

Although Dalton's atomism seemed to clarify one aspect of matter, it left open the long-standing question of why certain materials joined with others, while others (such as gold) did not combine. The general idea among chemists was that each material had a natural affinity for other substances and that this led to certain combinations and not others. There was never a single theory of affinity agreed to by all chemists, but a number of chemists attempted to chart affinity, and this was in some ways a prelude to the hunt for order among the elements that produced the periodic table. In addition to simple combination, it was clear that some materials would combine with many things, but there seemed to be a hierarchy of combination. The power of combination was so great between some substances that a strong combination would displace a weaker one in an existing compound. The idea of affinity alone as the basis for combination was dealt a blow in 1803 when Berthollet published his book *Essai de statique chimique.* He demonstrated that the amount of reacting material available also affected the creation of the final compound. This would make it difficult, if not impossible, to determine anything but simple (one-to-one) affinity. The problem of mass action would remain unresolved for another 60 years.

In 1800, a new tool for examining chemicals became available. This was the "voltaic pile," designed by Alessandro Volta (1774–1827). The voltaic pile (or what we today would call a battery) consisted of alternating layers of zinc and silver disks. The battery allowed for the production of a continuous flow of electricity or current, rather than the sudden discharge produced by Leyden jars. When researchers ran a current through water, they discovered that electricity decomposed the water into oxygen (which collected at the positive pole) and hydrogen (at the negative pole). Although this confirmed earlier work, it would become a clue for later chemists about the nature of bonds within molecules. This new branch of electrochemistry was particularly important for Humphry Davy (1778–1829), who used powerful batteries to decompose potash, isolating sodium and potassium in 1807. Over the next few years, he also showed that various alkaline earths were compounds, isolating the metals calcium, strontium, and barium from their oxides.

Further work on electrochemistry was done by the Swedish chemist Jöns Jakob Berzelius (1779–1848), who took Dalton's concept of atoms and combined it with the concept of electrical attraction. Since compounds such as water or metallic oxides could be separated by electrolysis, it seemed reasonable to assume that the elements had an electrochemical nature that accounted for their combination into compounds. Using this idea, Berzelius arranged the elements in a series from oxygen to potassium. While this was very useful for some compounds, it led him to claim that iodine and chlorine could not be elements but had to be oxides of as-yet undiscovered elements because they seemed to form electronegative salts. This problem was cleared up with further analytical work in the 1820s, and Berzelius created a separate category for iodine, chlorine, and bromine.

Although Dalton's atomism provided a strong foundation for some aspects of chemistry and helped to reinforce the new analytical chemistry championed by Lavoisier, it did not completely clarify the problem of the elements.

One of the major problems was volumetric relations. In 1808, Joseph Louis Gay-Lussac (1778–1850) had been working on the proportions of gas combination, noting, for example, that 100 volumes of nitrogen combined with 300 volumes of hydrogen to produce 200 volumes of ammonia. It seemed to follow from these experiments that combination took place in definite and simple ratios. The findings were rejected by Dalton, however, because they seemed to imply that atoms split, which was, by definition, impossible. This was most obvious in the case of water, where two volumes of hydrogen combined with one volume of oxygen to produce two volumes of water. If the combination of hydrogen and oxygen was simple (that is water was HO), then the atoms in the one volume of oxygen would have to split in two. The idea that oxygen might be a binary substance (what would be written today as O_2) seemed to be against reason, either because of Dalton's idea of repulsive force (which he based on Newtonian physics) or because of the difficulty of electro repulsion that prevented particles of like charge from combining.

In addition to the volume problem, there was a philosophical objection to Dalton's atomism. Under his system, there were at least 50 different elements. This seemed like a huge number to many chemists. The solution for many chemists was to make a distinction between physical atoms and chemical atoms. A physical atom in some ways harkened back to prime matter, since it was singular and did not exist as a separate entity in nature. The chemist worked with chemical atoms, which existed in nature. Under real-world conditions, chemical atoms could not be decomposed, so there was no contradiction of Lavoisier's definition of an element, and these elements could be classified by their characteristics, such as weight or electrochemical behavior. Atoms could be simple, but matter could be complex.

Chemistry had now become more organized and the language clearer, but significant problems remained. Chemists were still divided as to how atoms combined and whether elements could be used as a reliable basis for understanding the structure of matter. The famous organic chemist Friedrich August Kekulé (1829–1896), who determined the ring structure of benzene, noted in 1864 that there were at least 20 different formulas in print for acetic acid. Some of the formulas were different simply because the chemists were using different notation to represent the same thing, but some versions of the acetic acid formula were different because the chemists were using a different system of atomic relations.

By the time Kekulé was doing his work, the concept of the atom had largely replaced the more general conception of matter presented by the corpuscularians. Chemistry was established as both a discipline and a profession. The examination of matter had become more precise and systematic, and the problem of nomenclature and communication, if not completely solved, had been reformed to the point that chemists could talk without having to explain every name for a chemical, tool, or laboratory procedure. Alchemy had been relegated to the junk closet of history, increasingly regarded (if it was thought about at all) by chemists as a flawed philosophy that had nonetheless, if only because it created opposition, led to the creation of to the "real" science of chemistry.

Despite the increasing sophistication of chemists, the elements were still not understood. Many chemists simply accepted the existence of certain substances without inquiring too closely about their specific structure, preferring to work with what was observable. In the growing field of organic chemistry, it did not really matter whether carbon was an element or a compound, a chemical atom or a physical atom, so long as the products analyzed and synthesized were consistent. Toward the end of the nineteenth century, a number of physical chemists, led by Friedrich Wilhelm Ostwald (1853–1932), went so far as to reject material atoms altogether, creating a system of chemistry based on forces. These scientists, known as the Energeticists, argued that matter was made not of hard particles but rather of fields of gravity and electromagnetic forces that interacted. This view was also partly based on the recognition of the wave nature of light. Any substance, if heated sufficiently, produces light. Since it was difficult (although not impossible) to describe light waves using a particle model, it seemed logical to assume that a particle model of matter was wrong or at least deeply flawed. While the physical atom would be reinstated in chemistry in the twentieth century, it would be a much different object from Dalton's indivisible particles.

8

SEEKING ORDER: THE PERIODIC TABLE

One of the primary themes of matter theory over the generations has been making order out of what appeared to be disordered or chaotic. By 1860, the number of compounds known to chemists was growing rapidly and would continue to grow at an ever fast pace, particularly as organic chemistry began to introduce new products such as aniline dyes and synthetic drugs. Depending on the philosophical position of the chemist, the number of true elements was also growing. Although not all chemists felt that understanding the elements (if elements existed at all) would reveal the nature of complex compounds, most agreed that some system for describing and comparing substances was needed.

Chemistry at the beginning of the nineteenth century was in a position analogous to Ptolemaic astronomy. Around 100 C.E., Ptolemy created his astronomical system, which could be used to accurately chart the movement of all the visible planets, the sun, and the moon. His system worked perfectly well for navigation, time keeping, and all the other activities for which knowing the place of celestial objects was needed. Yet, Ptolemy's system had a philosophical problem. Each of the planets had its own system of movement, and the laws of planetary motion were different from those that governed motion on the Earth. Newton's astronomy and physics put all the planets, the stars, and motion here on Earth into a single system.

In chemistry, each element, like the Ptolemaic planets, seemed to have its own characteristics. Chemists had hoped for generations that there would be a unifying system for matter that would be the equivalent of Newton's physics, but if there was a unifying principle, it was proving difficult to determine. Even among the elements, there were huge apparent differences. The nature of the elements seemed so varied as to defy easy comparison. These were liquids, solids, and gases, metallic and nonmetallic elements, electropositive and negative elements, high-affinity and low-affinity elements. Some elements were combustible, whereas others resisted combustion; some occurred naturally, whereas others existed only if decomposed from other things. And these differences didn't even touch on the problem of physical composition, such as size and shape.

Jöns Berzelius, one of the leading analytical chemists of his age, focused on atomic weight as a way of ordering the elements. Between 1814 and 1826, he did thousands of tests to purify and measure the weight of elements. Establishing atomic weight was, however, not simply a matter of weighing some hydrogen or some gold. The weights had to be built up using a system of comparisons to an arbitrarily selected standard. All systems of weights and measures are arbitrary, whether it is using the length of a person's arm for length (the yard) or establishing that the mass of a cubic decimeter of water equals a kilogram.[1] Determining the weight (or, more precisely, the mass) of an individual atom was impossible by direct measurement, but, because chemicals combine in fixed proportions, it was possible to determine the difference in mass of different types of atoms. In turn, this would allow scientists to establish the theoretical mass of an individual atom. Dalton, for example, set hydrogen equal to 1. This made sense, since hydrogen seemed to be the lightest gas (and hence the lightest element). The weight measurement could then be based on the system:

$$HX = H + X$$
$$X = HX - H$$

where the unknown weight (X) is determined by subtracting the known weight of H (hydrogen) from the known weight of the combined compound (HX). Unfortunately, hydrogen does not combine with everything, so there are limits on this system. On the other hand, oxygen combines with many different elements, so a number of chemists used it as the basis for comparison. In Berzelius's system, he set the value of oxygen equal to 100.

In 1819, the French chemists Pierre Louis Dulong (1785–1838) and Alexis Thérèse Petit (1791–1820) reported that in solid elements, the product of specific heat and atomic weight equaled a constant. Specific heat was the amount of heat required to change a fixed mass of a substance by one degree in temperature. While Dulong and Petit hoped that their system would reveal atomic weights, the relationship was only approximate and was often difficult to determine. While it did not lead to a breakthrough in understanding atomic weight, it did offer a way to confirm the magnitude of results from other forms of analysis, and it did help to correct a number of problematic findings such as the high atomic weight of lead found by Berzelius.

Part of the reason that there was increasing pressure to find a unifying system was that the number of elements had almost doubled from the time of Lavoisier. Lavoisier had listed 33 elements, but by 1844 an additional 31 new elements had been added to the list. There was a slowing of discovery between 1844 and 1859, but two new elements were added when Robert Wilhelm Bunsen (1811–1899) and Gustav Robert Kirchhoff (1824–1887) announced the discovery of cesium in 1860, followed by rubidium in 1861, bring the total list of elements to 66. They had discovered the two new elements using spectroscopy, and named them after the bright color lines that Bunsen and Kirchhoff observed—blue for cesium and ruby red for rubidium. Although the spectral refraction of light had been known in ancient times, the use of spectra as a scientific tool really started with Joseph Fraunhofer (1787–1826), a Bavarian lens manufacturer, who in 1814 noted that there were dark and bright lines in the spectrum of light he observed from various flames and sunlight. He labeled the most distinct lines A through H and used their existence to study the refractive index of glass.

In 1855, Bunsen was using flame color as a way of identifying salts. He had a new laboratory, which was equipped with coal gas (a new innovation used for street lighting in the city of Heidelberg), replacing the alcohol and oil lamps used for heating in laboratory experiments. Since commercially available gas burners were designed to produce light rather than heat, Bunsen worked with the university mechanic, Peter Desaga, to design a burner that would produce high temperatures and a colorless flame. The result was the Bunsen burner, and it became a common tool in laboratories around the world.

Kirchhoff was a physicist, and he suggested to Bunsen that he might get better results from his color studies if he used prisms rather than colored glass to filter out the flame colors made by the salts he was looking at. Together, they constructed a spectroscope, which consisted of Bunsen's burner, telescopic sights, and a prism. Light from any source that emitted light had an individual and characteristic pattern of colored lines. Thus glowing hot iron produced a set of lines different from those produced by incandescent aluminum.

Kirchhoff worked out the laws of spectroscopy and in 1859 published his findings, which included three laws:

1. An incandescent body gives off light in a continuous spectrum.
2. An excited (heated) body gives off a bright-line spectrum.
3. A vapor of an element absorbs white light in the places of the spectrum (producing dark lines) at the same points where the element heated to incandescence produce bright lines.

Kirchhoff and Bunsen's work created a new analytical tool for chemists, and, in part because of the introduction of the spectroscope, thallium was identified as an element, in 1861, and indium, in 1863. The tool was also used by astronomers, who used Kirchhoff's rules to determine the chemical composition of stars. In the long run, it was shown that all the heavy elements were by-products of stellar fusion. Even the atoms of our bodies were made from the stuff of stars.

As more tools and techniques were becoming available to chemists, there was greater confusion, rather than less, about the underlying principles of matter. A number of scientists attempted to bring order to chemical relations by grouping elements together in various ways. One of the earliest attempts to do

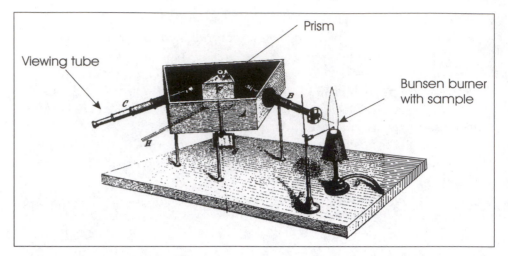

Figure 15: Gustav Kirchhoff and Robert Bunsen's spectrocope, from "Chemical analysis by Observation of Spectra," (1860).

more than measure atomic weight was that of Johan Döbereiner (1780–1849). He used the atomic weights published by Berzelius but noted that some elements seemed to form groups of three, linked by atomic weight and characteristics. For example, he found that the atomic weight of sodium was the mean of the weight of lithium (lighter) and potassium (heavier). All three seemed to have similar chemical activity. He called these groups triads and presented a number of complete and incomplete sets.

Examples of Döbereiner's triad system

Calcium	Chlorine	Sulfur	Manganese
Strontium	Bromine	Selenium	Chromium
Barium	Iodine	Tellurium	Iron

Incomplete triads

Boron	Beryllium
?	?
Silicon	Aluminum

Döbereiner predicted that the missing elements in his incomplete triads would eventually be found. While Döbereiner's insight was interesting, it suffered from a number of problems. First, he could not include all the elements; he left the important elements nitrogen, oxygen, carbon, and hydrogen out of the system, claiming them to be isolated. Second, the atomic weight relations were only approximate and were constantly being revised, leaving the triads open to revision resulting from new experimental data. Although Döbereiner predicted the approximate atomic weight of bromine before it was determined by analysis, his system was not widely adopted. The necessity of separate classes of some chemicals also seemed to run contrary to the idea of a single, all-encompassing system. Döbereiner might have been on the right track, but his work seemed more Ptolemaic than Newtonian, relying on separate systems for each group rather than proposing a unified system for all the elements.

Another intriguing but failed attempt to organize the elements was presented by A. E. Béguyer de Chancourtois (1820–1886), a professor of geology. Using what he called the "telluric helix," he arranged the elements by weight on a graph around a cylinder. Using 16 columns (setting hydrogen to 1 and oxygen to 16), he placed the other elements around the cylinder in descending order at a 45° angle. He found that there were close relationships between many of the elements that lined up vertically. De Chancourtois's work might have gained wider notice when he reported his idea to the Académie des Sciences in 1862, but his writing was unclear and he did not include a diagram of the telluric helix.

In 1863, John Newlands (1837–1898) created a table of elements grouped by atomic weight and by chemical characteristic. Newlands noticed that in an eight-column table, the first and eighth element in a row resembled each other. This relationship then repeated, so that the fifteenth resembled the first and eighth. He called this the "law of octaves," borrowing the idea from musical scales.

Newland's Law of Octaves. Elements.
1/8/15 shared characteristics

			[1]			
2	3	4	5	6	7	[8]
9	10	11	12	13	14	[15]

He also numbered the elements. In his early work, he left spaces for what he assumed were undiscovered elements that would fill in the octaves or family groups on his chart. The refined idea was presented to the Chemical Society in London in 1866, but Newlands, for reasons that are not completely clear, decided to remove the spaces for undiscovered elements. This caused problems with the chemical relations within some of the groups and made it look as if the table were not based on any consistent system. The new table was heavily criticized, with one member of the Chemical Society asking if Newlands had thought of simply arranging the elements alphabetically. Although Newlands had come very close to establishing the periodic nature of the elements, when his paper was rejected for publication by the *Journal of the Chemical Society*, he turned to other work.

The creation of the modern periodic table of elements was the result of very similar insights by two men, Dimitri Ivanovich Mendeleev (1834–1907)[2] and Julius Lothar Meyer (1830–1895). In terms of priority, Meyer presented a partial table of elements arranged by atomic weight in his widely read book *Die modernen Theorien der Chemie* (1864). It was written in part as a presentation of the ideas of Amedeo Avogadro (1776–1856) as revived by Stanislao Cannizzaro (1826–1910). Avogadro's hypothesis was that equal volumes of all gases, at the same temperature and pressure, contain the same number of molecules. Given this fact, it was possible to determine the molecular formula for gases and from that to establish their atomic weight. Avogadro's theory offered a solution to a problem that had been plaguing matter theorists with regard to the products of combined gases, and, by extension, all compounds.

Avogadro's hypothesis was one of the great ideas of modern chemistry, but when it was first proposed, it was largely ignored. Originally trained as a lawyer, Avogadro spent most of his professional life researching and teaching physics. He knew that matter theory as presented by scientists such as Dalton was based on a concept of fixed proportions for the creation of compounds from simpler elements, but, when experiments were conducted, the volume of material produced did not match what was predicted by the theory. For example:

Predicted

2 volumes of hydrogen + 1 volume of oxygen = 1 volume steam (water vapor)

Observed

2 volumes of hydrogen + 1 volume of oxygen = 2 volume steam

Avogadro's solution was to propose the existence of diatomic elements, or particles that consisted of two atoms of the same material. It followed that the number of particles in the two volumes of hydrogen would be equal to the number of particles in one volume of oxygen, if the oxygen particles consisted of two atoms of oxygen. Once the actual number of atoms was determined, accurate comparisons of atomic weight could be achieved. In Avogadro's words:

> M. Gay-Lussac has shown in an interesting Memoir that gases always unite in a very simple proportion by volume, and that when the result of the union is a gas, its volume also is very simply related to those of its components. But the quantitative proportions of substances in compounds seem only to depend on the relative number of molecules which combine, and on the number of composite molecules which result. It must then be admitted that very simple relations also exist between the volumes of gaseous substances and the numbers of simple or compound molecules which form them. The first hypothesis to present itself in this connection, and apparently even the only admissible one, is the supposition that the number of integral molecules in any gases is always the same for equal volumes, or always proportional to the volumes.[3]

Avogadro's 1811 work had largely been ignored during his lifetime, in part because it did not fit with the existing theories of elements. It seemed to require either the splitting of elements (which by definition could not be divided) or the existence of some bi-atomic molecules, which seemed awkward. Although Avogadro used Dalton's atomic theory and Gay-Lussac's law of combining volumes, he avoided the term "atom," referring to the smallest particles as "half molecules." When continued work based on Dalton's and Gay-Lussac's ideas failed to account for the discrepancy between theory and observation, the chemical community was more willing to consider Avogadro's hypothesis. That, combined with advances in analytical chemistry that gave more certainty to volumetric observation, opened the door for Cannizzaro, and when he presented Avogadro's ideas at the important Karlsruhe Conference in 1860, it quickly became one of the foundational ideas of modern chemistry.

In turn, Meyer's 1864 table of elements was a presentation of Avogadro's hypothesis as it applied to the known elements. Although Meyer would continue to work on his table of elements, seeking to group them by weight and characteristics, it was Mendeleev who produced the first widely accepted periodic table of elements. Part of the power of Mendeleev's work came from the almost instant recognition among chemists that the table really did put things into groups by related characteristics. The utility of the theoretical foundation of the periodic table also contributed to its swift acceptance. It was not just that the table brought order to the apparent chaos of the known elements but also that it allowed Mendeleev to predict the existence *and characteristics* of as-yet undiscovered elements.

Dimitri Mendeleev was the youngest children of at least 14 children.[4] His father was a schoolteacher in Tobolsk, Siberia, but, when illness made him blind, he was forced to retire. This placed a great strain on the family, since the pension he received was meager, so Mendeleev's mother, Maria, ran a glass-making factory to provide an income. Mendeleev's early schooling was heavily oriented toward classical education; he learned Greek and Latin, and in these he was not a particularly outstanding student. His family, particularly his mother, believed strongly in education, even opening a school for children of the workers at the glass factory, so he also received private tutoring in science. He showed great aptitude for all aspects of science.

The Mendeleev family then suffered a series of catastrophes. First, Mendeleev's father died in 1847. The following year, the glass factory burned to the ground. Maria decided that she would take her two remaining dependent children, Dimitri and his sister, Liza, to Moscow so that Dimitri could get higher education. The three made the 1,300-mile journey on foot and in wagons when they could find rides. In Moscow, the Mendeleevs' bad luck continued. Entry to the university was based on a quota system, but Siberia had not yet been assigned a quota, so Dimitri was not even considered for admission. Other schools in Moscow simply turned down his applications because his Siberian education was not seen as adequate.

In desperation, the Mendeleevs traveled a further 400 miles to St. Petersburg, the capital of Russia. In St. Petersburg, the answer was the same, but fortunately the head of the Central Pedagogical Institute, a training college for high school teachers (where Mendeleev's father had studied) was an old friend of

Dimitri's father. Finally, Dimitri was admitted for higher education and was even granted a small scholarship to cover living expenses.

Their luck had not turned, however. Within three months, Maria Mendeleev had died, and about a year later, Liza also died. Mendeleev was alone, but that was not the end of his trouble. A year after the death of Liza, Mendeleev started to bleed from his throat and ended up in hospital, where tuberculosis was diagnosed. He was given only a few months to live. Despite all these calamities, Mendeleev continued to study, and when he could, he did experiments in the Institute laboratories, even publishing several papers on his research. When Mendeleev graduated, in 1855, he won a gold medal as best student and was appointed to a job teaching science at the gymnasium (or academic high school) in Simferopol, in the Crimea. Mendeleev arrived at his job in the middle of the Crimean War, finding the school closed and no employment available. The situation took a turn for the better, however, when a surgeon diagnosed Mendeleev's disease as nonfatal. With new hope for the future, Mendeleev made his way back to St. Petersburg, where he found work at the University of St. Petersburg and continued his studies.

In 1859, Mendeleev received money from the government to study chemistry abroad. He traveled to Paris, where he studied with Henri Regnault, a brilliant experimentalist. He then went to Heidelberg, where he worked with Kirchhoff and Bunsen and was introduced to spectroscopic analysis. In 1860, Mendeleev attended the Karlsruhe Congress and heard Cannizzaro talk about Avogadro's hypothesis and the solution to the volume/chemical combination problem. When Mendeleev returned to St. Petersburg, he was fired with enthusiasm for chemistry and convinced that he could contribute important work. One of his self-appointed tasks was to modernize chemical education in Russia. He developed into a charismatic instructor and brought great powers of concentration to his work. For example, no Russian-language textbook on modern organic chemistry existed, so he produced a 500-page book on organic chemistry in just two months.

Mendeleev grew increasingly concerned that without a central organizing system for chemistry, the subject would remain a collection of ad hoc theories and bits of technical methodology. In 1869, while still thinking about the elements, he completed the first volume of his magnum opus, *The Principles of Chemistry*. When starting the second volume, Mendeleev was forced to look for some unifying characteristics in the elements that could be used for the unifying structure of the next section of the book. From his own recounting of the story and information from friends, we know the time Mendeleev found his structure. On Friday, February 14, 1869, he was in the midst of writing. He wanted to get things straightened out over the weekend, because he was going away on an inspection tour of farms in Tver on the following Monday.

Mendeleev started by writing out the elements in various patterns. Eventually, he prepared a set of cards on which he wrote the properties and characteristics of all the known elements, one card to an element. Then, playing a kind of scientific solitaire, he arranged and rearranged these cards. When he arranged them by atomic weight, he recognized that certain characteristics repeated as weight went up. There was something in the patterns, but he could not see it. Monday came and went, and he still had not found what he was looking for. Exhausted, he fell asleep at his desk and dreamed of a table of elements. When he awoke, he wrote down what he remembered from his dream. Two weeks later, he published "The Relation between the Properties and Atomic Weights of the Elements" in the *Journal of the Russian Chemical Society* (1869).

The table presented the elements in ascending order according to weight, using a system of horizontal rows (called "series") and vertical columns (called "groups"). In his revised and improved table of 1871, there were 12 series divided into 8 columns, which he labeled Group I through VIII. Within the groups, the elements also ascended from lightest at the top to heaviest at the bottom, but all the elements in the group shared common properties. When there was no element that fit the pattern, he left the space in the table blank to indicate a missing element.

Mendeleev was convinced that he had discovered what he called the Periodic Law. The principle of the law was that the characteristics of the elements would vary periodically (that is, repeat at set intervals) as atomic weight went up. Characteristics such as specific density, oxidation states, and affinity (degree of chemical interactions) would vary for each element, but such variation was within a specific range that was common to a particular group. Thus, calcium (element 20) might be much heavier than magnesium (element 12) and only a bit heavier than potassium (element 19), but calcium and magnesium were related by chemical behavior.

Even though Mendeleev's table worked brilliantly for many of the elements, there were problems, and this led to controversy over the accuracy and utility of his system. The most obvious one was that the table was not completely consistent. While the characteristics of the elements in the groups seemed to be common, on closer examination, there were anomalies. Some of the elements

Reihen	Gruppe I. — R^2O	Gruppe II. — RO	Gruppe III. —. R^2O^3	Gruppe IV. RH^4 RO^2	Gruppe V. RH^3 R^2O^5	Gruppe VI. RH^2 RO^3	Gruppe VII. RH R^2O^7	Gruppe VIII. — RO^4
1	H＝1							
2	Li＝7	Be＝9,4	B＝11	C＝12	N＝14	O＝16	F＝19	
3	Na＝23	Mg＝24	Al＝27,3	Si＝28	P＝31	S＝32	Cl＝35,5	
4	K＝39	Ca＝40	—＝44	Ti＝48	V＝51	Cr＝52	Mn＝55	Fe＝56, Co＝59, Ni＝59, Cu＝63.
5	(Cu＝63)	Zn＝65	—＝68	—＝72	As＝75	Se＝78	Br＝80	
6	Rb＝85	Sr＝87	?Yt＝88	Zr＝90	Nb＝94	Mo＝96	—＝100	Ru＝104, Rh＝104, Pd＝106, Ag＝108.
7	(Ag＝108)	Cd＝112	In＝113	Sn＝118	Sb＝122	Te＝125	J＝127	
8	Cs＝133	Ba＝137	?Di＝138	?Ce＝140	—	—	—	— — — —
9	(—)	—	—	—	—	—	—	
10	—	—	?Er＝178	?La＝180	Ta＝182	W＝184	—	Os＝195, Ir＝197, Pt＝198, Au＝199.
11	(Au＝199)	Hg＝200	Tl＝204	Pb＝207	Bi＝208	—	—	
12	—	—	—	Th＝231	—	U＝240	—	— — — —

Figure 16: Mendeleev's periodic table from *Annalen der Chemie,* (1872).

ended up grouped with others that seemed out of order if considered by atomic weight. To make the element groups work by characteristic, Mendeleev asserted that some of the atomic weights must have been miscalculated. In other words, the experimental chemists had made errors and would have to go back to their laboratories and come up with better results. This kind of bold assertion in science was risky business. Mendeleev was right, and his system was strengthened. If he had been wrong, he would have been little more than a footnote in the history of chemistry.

There was also a problem with affinity. Affinity was an attempt to measure the degree that elements would join together, and some of the groups of elements seemed to have different levels of affinity. This problem would be resolved with the introduction of the concept of valency, which was a measure of the specific number of atoms that could be linked to an atom of an element. Like his assertions about the problem with some of the atomic weight calculations, Mendeleev assumed that future discoveries would prove his system correct.

What helped make Mendeleev's system seem more plausible was the coincidental publication of Lothar Meyer's periodic table. Meyer had attended the 1860 Karlsruhe conference just as Mendeleev had, and, like Mendeleev, he had gone away convinced that there was some pattern that united the elements through atomic weights. He had sketched out an incomplete table of elements in his book *Die modernen Theorien der Chemie* (1864), and in 1868 he worked out a much more complete table. He did not publish this until 1870, by which time he was fully aware of Mendeleev's work. In addition to providing independent confirmation of Mendeleev's system, Meyer also added a supporting observation, showing that there was a periodic relationship between atomic number and atomic volume. Atomic volume represents the volume occupied by one mole of an element in its solid state. Meyer showed that atomic volume equaled atomic weight divided by the density of the solid.

One of the most controversial aspects of Mendeleev's idea was the blank spaces he left in his table. From the characteristics of the groups and the pattern of those known elements around the blank spaces, Mendeleev made a number of predictions about the "missing" elements. These included atomic weight and specific gravity (now known as relative density), which measures the density of a material compared to an established baseline. For solids or liquids, the material is compared to the density of water at 4° C, oxide properties, appearance, and even color.

The most extraordinary of the predictions to be demonstrated was the discovery, in 1875, of gallium. The French chemist Paul Lecoq de Boisbaudran (1838–1912) announced the discovery of the new element, which he named "gallium" after the Latin name for France, in a letter to the Académie des Sciences.[5] Lecoq was unaware of Mendeleev's work on the periodic law at the time, but when Mendeleev read about Lecoq's work, he felt that it was proof of his system. This led to a misunderstanding between Lecoq and Mendeleev, since it seemed at first that Mendeleev was claiming priority of discovery.

Mendeleev's predicted element and date	Discovered element and date
coronium (later newtonium) 1902	No element
dvi-caesium 1871	francium 1939
dvi-tellurium 1889	polonium 1898
eka-aluminium 1871	gallium 1875
eka-boron 1871	scandium 1879
eka-caesium 1871	No element
eka-manganese 1871	technetium 1939
eka-niobium 1871	No element
eka-silicon 1871	germanium 1886
eka-tantalum 1871	protactinium 1871
Ether 1902	No element
tri-manganese 1871	rhenium 1925

The priority issue was later straightened out, but there remained a discrepancy between Mendeleev's prediction and Lecoq's initial figure for one of the properties of gallium. Mendeleev had predicted that the specific gravity of the element would be 5.9, while Lecoq found it to be 4.7. Mendeleev predicted that gallium would be almost six times as dense as water, while Lecoq found that it was less than five times as dense. Mendeleev wrote to Lecoq suggesting that he repeat the experiment with a larger sample. When this was done, Lecoq found that the specific gravity was, as predicted, 5.9. It seemed a clear demonstration that the periodic table really represented the hidden organization of the elements.

Further confirmation came when Lars Nilson (1840–1899) announced the discovery of scandium (after Scandinavia), in 1879, which matched Mendeleev's eka-boron, and Clemens Winkler's (1838–1904) announcement of the discovery of germanium (after his home country of Germany) in 1886 which matched eka-silicon. Mendeleev's theoretical position triumphed, and his table began to be accepted as a genuine insight into the order of the material world.

Results like these were largely responsible for making Mendeleev famous even though his system and Meyer's were almost identical. Meyer was far more reluctant to make predictions about unknown elements, preferring to present his work more for its value in resolving the issue of Avogadro's hypothesis and the issue of atomic weight. Mendeleev's story of dreams, missed trains, and maniacal work also garnered him a level of public interest that Meyer's careful, almost clerical approach could not match.

Despite the apparent power of Mendeleev's system, it was not completely reliable. Eka-niobium and eka-caesium were eventually shown not to exist, while a number of his predictions, while accurate for place in the table, were not as accurate in regard to their details. The main problems were with the rare-earth elements. The first of the rare-earths identified was yttria (after the village of Ytterby, near Stockholm). It was discovered in 1797. These elements tended to be very scarce and heavy and to have properties that were difficult to measure. Numerous new "elements" were discovered and then later found to be either re-discoveries with slightly different measurements or not really elements but hard-to-separate compounds. Mendeleev placed the six known rare-earth elements into his table, even though it led to a huge number of gaps, or missing elements. Although one of the six elements would later disappear (it turned out to be praseodymium and neodymium), by 1886 six new rare-earth elements had been identified, complicating the issue even further. Bohuslav Brauner (1855–1935), one of Mendeleev's supporters, suggested, in 1902, that the rare-earth elements be placed in a separate list, fitting into one space in Group IV between lanthanum and tantalum. Mendeleev would not accept the idea, arguing that the periodic law required each element to be given its own place, regardless of what it did to the ordering of the rest of the table.

Mendeleev also ran into problems with his characterization of coronium (which he later renamed newtonium after the great scientist) and ether. Mendeleev had used his periodic rule to work upward through the atomic weights of the elements, but it seemed reasonable to go down as well, and that led Mendeleev and others to speculate that there might be elements lighter than hydrogen.

Although Mendeleev speculated about the possibility of gases lighter than hydrogen, it was an effort to clarify the characteristics of two well-known gases that led to a major reform of the periodic table. The origin of this story went all the way back to Henry Cavendish, who in 1783 had sent sparks through mixtures of dephlogisticated air (oxygen) and phlogisticated air (nitrogen). Most of the materials produced were oxides of nitrogen collected as nitrous acid, but a tiny bit was left over. Cavendish said it was not more than 1/120 of the product, but he did not study the fraction, probably assuming it was the result of contaminants or experimental error. In 1888, Lord Rayleigh, working at the Cavendish laboratory, decided to recalculate the densities of nitrogen and oxygen. Using very careful methods and sensitive measuring equipment, he found that there was a slight difference in the weight of atmospheric nitrogen and "chemical" or laboratory-produced nitrogen, which was a tiny bit lighter. Until the discrepancy was cleared up, the density figures were in doubt. The chemist William Ramsay (1852–1916), working at University College, London, looked at the results and was also unable to account for the difference.

By 1894, Ramsay and Rayleigh believed that the discrepancy was caused by an unknown inert gas. They agreed to work together on the problem and produced a small sample. Rayleigh investigated its physical properties and Ramsay, its chemical. Ramsay was unable to get the gas to combine with any other substance, so they proposed the name "argon," from the Greek *argos*, meaning inactive or idle. There were strong objections to their speculation about an unknown gas, since atmospheric air had been so extensively studied, but with the use of the spectroscope, the conclusion was inescapable, and in 1895 the existence of argon was publicly announced.

Argon was an amazing discovery, but it created a new set of questions. It seemed that there should be other gases with densities that could be predicated by the period law. The weight of argon at 40, however, suggested to Mendeleev and others that there was something wrong with the process, because calcium had an atomic weight of 40. They argued that argon had to be diatomic, so its real weight should be 20. Ramsay and Rayleigh argued that argon was monoatomic. The issue was made more complicated when, in 1895, Ramsay isolated helium. Spectroscopic analysis showed that this was the same element that had been identified in 1868 by Joseph Lockyer (1836–1920) while observing the solar spectrum. Rather than one inert or zero-valence gas, there were now two.

By 1898, Ramsay and his assistant Morris Travers (1872–1961) had discovered neon, krypton, and xenon. All of them were chemically inert, so they were called the "noble" gases (they would not mix with "common" elements), and also they could be found in trace amounts in atmospheric air. In terms of the periodic table, this meant that a new column was needed, but, rather than disrupt the whole system, it actually confirmed the utility of the periodic system. Ramsay had predicted the characteristics of the new elements, and their characteristics had come out close to the expected values in each case. The last element of the column was added in 1903 with the discovery of the radioactive element radon by Frederick Soddy. In 1904, Ramsay won the Nobel Prize for chemistry for his work.

Despite the work of Ramsay and Rayleigh on the noble gases, Mendeleev continued to speculate about lighter elements. In 1902, he predicted that if hydrogen had an atomic weight of 1, then that of coronium would be 0.4 and that of ether 0.17. While these predictions made perfect sense based on the periodic law, they were actually based on a conception of physics that was on the verge of being overturned.

In a sense, the problem with Mendeleev's view of the rare earths and his predictions about coronium and ether were related to the same issue: the new physics. The heavy rare-earth elements, which are today generally grouped in a separate subtable, could not be understood until the physics and structure of atoms were revealed by new work being done on the structure of the atom itself. At the other end of the table, the very idea that coronium and ether were necessary parts of the material universe was being taken apart by Einstein and the theory of relativity. Although Mendeleev's insight into the hidden order in matter was brilliant, it was limited to a particular segment of the material world and reflected the understanding of physics and chemistry of the era. Rather than becoming a universal law, the periodic table became a useful tool. On one level, it became a cornerstone for education and communication, and it really did bring order to the chaos of the chemical world, but was at another level it was only the first step in understanding the hidden order of the elements.

The introduction of the periodic table and the periodic law that allowed elements to be grouped gave scientists a new vision of the structure of the universe. Even the material of the stars was revealed by the work that came from

Mendeleev's powerful insight. For many chemists, the periodic table was the last theoretical tool they needed, since the table made clear the framework of matter. There would be much more work done refining and adding data to the table over the next century, but the basic principles were set. The discovery of the "missing" elements and the addition of the noble gases confirmed the truth of the periodic law and the utility of the table. John Newlands, whose work had identified many of the periodic properties of the elements, was eventually awarded the Davy Medal by the Royal Society in 1887, and, in 1998, the Royal Society of Chemistry unveiled a plaque at his birthplace acknowledging his discovery of the periodic law.

For Mendeleev and most chemists, the elements were individual entities and the smallest component of chemical reality. For the next generation of matter theorists—the new physicists and chemists in the post-Einstein world—elements were specific combinations of smaller units. The majority of the periodic table was made up of islands of stability in a sea of theoretical possibilities. The very stability of what an element was began to disappear in the upper reaches of the periodic table. A new way of seeing the elements was being born.

NOTES

1. The original mass of water definition of the kilogram was created in 1799. It was replaced in 1901 when a specific artifact made of 90 percent platinum and 10 percent iridium alloy was created by the International Bureau of Weights and Measures as the standard for the kilogram. The kilogram is the only remaining physical unit of measurement that is defined by an actual object.

2. The spelling of Mendeleev's name is often confusing, having been translated from the Cyrillic alphabet. At least four spellings are in use: Mendeleyev, Mendeleev, Mendeléeff, and Mendelejeff. The most common version in English is Mendeleev.

3. Lorenzo Romano Amadeo Carlo Avogadro, "Essay on a Manner of Dermining the Relative Masses of the Elementary Molecules of Bodies, and the Proportions in Which They Enter Into These Compounds," *Journal de Physique, de Chimie et d'Histoire naturelle,* vol. 73, (1811): 58. Translation from *Alembic Club Reprints,* No. 4, "Foundations of the Molecular Theory: Comprising Papers and Extracts by John Dalton, Joseph Louis Gay-Lussac, and Amadeo Avogadro, (1808–1811)," available at dbhs.wvusd.k12.ca.us/webdocs/Chem-History/Avogadro.html.

4. It is unclear whether there were 14 or 17 children.

5. Some of Paul Lecoq de Boisbaudran's contemporaries suggested that he really named the new element after himself, since *"lecoq"* ("cockerel") is *"gallus"* in Latin. Boisbaudran denied this charge.

9

THE ATOMIC ELEMENTS

Mendeleev's system was a breakthrough both technically and philosophically. It put together the materials of the world into a coherent package and facilitated the communication of chemical ideas by clarifying nomenclature as well as by unifying ideas about atoms and molecules. The periodic table of elements helped turned chemistry into an empirical science, or a science based on unified and uniform rules that could be demonstrated by experiment. Although specific aspects of chemistry (as opposed to alchemy) had historically been empirical because they were based on experiments, as a whole, the principles of chemical activity could not be tested until the objects of study were brought together. Without that unity, researchers were free to create new explanations for any experimental result that did not accord with other results. Things like the celestial ether, various imponderable fluids, and vital spirits had been created as unique substances to fill perceived gaps in matter theory, but they withered away when no place could be found for them within the periodic system.

And then the entire structure came under attack by a group of physicists and physical chemists who rejected the very idea of material atoms. These scientists, called the Energeticists, argued that everything was made up of fields of force. This new physics would undermine the foundation of the periodic table. It wouldn't exactly make elements disappear, but it would certainly transform them from physical objects into a collection of characteristics.

The resolution of this challenge came from two different sources that eventually converged to produce a greater understanding of the structure of matter and confirmed the utility of the periodic table. The first development was the investigation of certain oxide compounds of the rare-earth elements. These substances were, as their name implies, extremely scarce, and the search for them was one of the great projects in chemistry. These materials would eventually be the key to radioactivity. The second development was the effort to determine the physical reality of atoms and molecules. This led to a powerful convergence of experimental and theoretic evidence.

When Lavoisier reformed chemistry in France, his work was being introduced in Germany by Martin Heinrich Klaproth (1743–1817). Trained as an apothecary, Klaproth taught himself the new chemical philosophy and became an assistant to Valentin Rose (1736–1771), one of the leading chemists of the day. He became director of Rose's pharmaceutical laboratory after Rose's death. Klaproth was an extremely good and extremely exact chemist, and a significant part of his work was directed at analytical chemistry, particularly the determination of characteristics of unknown compounds. During his years as director, he discovered or verified the discovery of zirconium, uranium, tellurium, and titanium.

Starting in the late eighteenth century, a number of new elements were discovered, but a brief note about the process is necessary. Klaproth and other chemists often did not isolate the pure element but rather an oxide form. Most of the characteristics of the actual element could be determined from tests on the oxide compound, and in general the credit for the discovery of the element has been given to the scientist who first demonstrated its existence, rather than the one who isolated the pure form. That means that, in a number of cases, there are two dates for the discovery of an element. The first represents the identification of a unique element, whose characteristics can be determined from the oxide form, and the second is the date of isolation, when the pure element was created. Thus, Klaproth identified zirconium in 1789, but it was not isolated until 1824. Also, because of the complexity of the tests and the very close range of characteristics, a number of rare-earth elements were "discovered" and then later proven to be compounds or previously identified. See Appendix Two for the history of the individual elements.

One town stands out in the history of the elements—Ytterby, near Stockholm. Minerals discovered at Ytterby were found to contain unknown substances that eventually added to the list of elements and forced chemists to improve their techniques as they hunted for new elements with complex characteristics in vanishingly small samples. Since the possibility of finding a new element was a major challenge and offered the chance of scientific immortality, several chemists were willing to do the hard and exacting work necessary. Starting in 1794, John Gadolin (1760–1852) discovered a black mineral with previously unknown characteristics. In 1797, A.G. Ekeberg argued that it was a new element and named it "yttria." Klaproth identified a different rare-earth, which he called "ceria." He handed on the research to Carl Mosander (1797–1858), who in turn identified several new elements from the original rare-earth element. From Gadolin's yttria and Klaproth's ceria, chemists would eventually isolate 17 elements.

During the same period that the rare earths were being investigated, a smaller but still analytically difficult family of elements was discovered. Platinum had been discovered around 1750 and had received a great deal of attention. It was extremely inert and seemed to have odd physical and chemical properties.

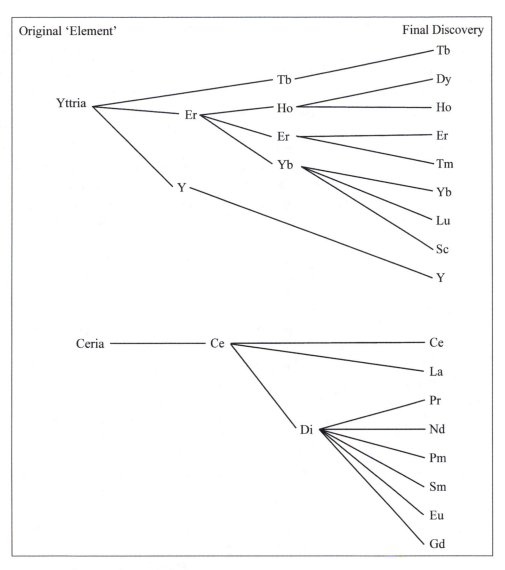

Figure 17: Ceria and yttria isolation.

William Hyde Wollaston (1766–1828) worked out a method of making malleable platinum metal, which he kept secret, selling the product commercially with great financial success. At the heart of the secret process was a system to purify the platinum and to amalgamate the resulting platinum powder by compression and hammering it. It was from the impurities that Wollaston identified palladium and rhodium, in 1803. A close friend of Wollaston, Smithson Tennant (1761–1815), identified two further platinum group elements, osmium and iridium, in 1804. The platinum-group element ruthenium was later discovered by Karl Karlovich Klaus (1796–1864), in 1844.

The yttria elements and, to a lesser extent, the platinum group elements both helped and confused the issues surrounding the periodic table of elements. When Mendeleev put together his table, the general groups of the elements fit well, but the rare-earth elements did not easily fit individually. In 1902, it was suggested that the rare-earth elements be placed in their own separate group and given a single space on the table between lanthanum and tantalum. Mendeleev had objected to this idea, and it was not until 1913 that the work of Henry Moseley (1887–1915) led to the acceptance of a separate box for the 14 known rare-earth elements.

Moseley's work was based on X-ray diffraction studies that revealed an even more powerful proof of the unity of the elements than Mendeleev's table of chemical characteristics. Moseley had worked with Ernest Rutherford (1871–1937) on radioactivity, but in 1912 he decided to use X-ray diffraction to examine the characteristics of the elements. He received training from Lawrence Bragg (1890–1971), the world's leading X-ray diffraction expert, and in 1913 began charting the X-ray spectra of the metallic elements.

In his experiments, he beamed X-rays at a sample of an element and then looked at the angles the rays made when they passed through a crystal of potassium ferrocyanide. As when light is passed through a prism, the angles created by X-rays could be used to determine the wavelengths and frequencies of the X-rays. Moseley found that the series he produced changed uniformly from element to element moving up the table from aluminum to gold. When he graphed the square root of the frequency against the number representing the element's position on the periodic table, he got a straight line. He called this the atomic number of the element, which was later shown to represent the positive charge in the nucleus or, in simpler terms, the number of protons. And since the number of electrons in a regular atom equals the number of protons, the electron content was also revealed.

By arranging the elements by their atomic number, rather than by their atomic weight, Moseley resolved the few problems that had persisted in Mendeleev's system. He also was then able to predict the existence of elements 43, 61, 72, 75, 87, and 91. All were later found. Further, because of the regularity of the wavelength studies, Moseley's work also experimentally demonstrated that there could be no new elements among the existing ones. The great confusion of the rare-earth elements was resolved, although because of their close relationships they were grouped together in a separate box. This helped to keep the other elements on the table in visually neater groups.

Moseley was looking for element 72 when World War I began. He enlisted in the British Army and was killed at Gallipoli in 1915, cutting short a career that would almost certainly have earned him a Nobel Prize.

A large part of the success of the hunt for new elements came because of the introduction of new tools such as the spectroscope and electrochemical apparatus. New ideas about the physics of heat also transformed the way scientists thought about materials, as the older fluid models (phlogiston and caloric) were replaced by a kinetic theory of heat. In the kinetic theory, it was the motion of the particles that determined the amount of heat in a substance. As research on electricity, matter, and light (spectra) continued, it became clear that the Daltonian atom model had problems. While it was generally

Moseley's Table of X-ray Spectra. Note the Regularity of Q_k and N Atomic Number.

	alpha line lambda $\times 10^8$ cm	Qk	N (atomic number)	beta line lambda $\times 10^8$ cm
Aluminum	8.364	12.05	13	7.912
Silicon	7.142	13.04	14	6.729
Chlorine	4.750	16.00	17	———
Potassium	3.759	17.98	19	3.463
Calcium	3.368	19.00	20	3.094
Titanium	2.758	20.99	22	2.524
Vanadium	2.519	21.96	23	2.297
Chromium	2.301	22.98	24	2.093
Manganese	2.111	23.99	25	1.818
Iron	1.946	24.99	26	1.765
Cobalt	1.798	26.00	27	1.629
Nickel	1.662	27.04	28	1.506
Copper	1.549	28.01	29	1.402
Zinc	1.445	29.01	30	1.306
Yttrium	0.838	38.1	39	———
Zirconium	0.794	39.1	40	———
Niobium	0.750	40.2	41	———
Molybdenum	0.721	41.2	42	———
Ruthenium	0.638	43.6	44	———
Palladium	0.584	45.6	46	———
Silver	0.560	46.6	47	———

Source: H. G. J. Moseley, "The High Frequency Spectra of the Elements," *Philosophical Magazine* 26 (1913): 1024

a reasonable model for organic chemistry and the most basic chemical expla-nations, it could not account for the increasing range of physical phenomena that were coming from the laboratories around the world. In particular, work with electricity was a problem. Electricity had gone from being a mysterious, largely uncontrollable substance in the eighteenth century to being a com-mercial product by the end of the nineteenth century. It was closely associated with matter but did not seem to be matter.

The key to the problem lay in the construction of a new instrument. In the 1850s, Heinrich Geissler (1814–1879), a scientific instrument maker and expert glassblower, created electrical discharge tubes. These were originally sealed glass tubes with a cathode at one end and an anode at the other. They were the precursor of many electronic devices ranging from the neon light tube to the transistor. The tubes needed a high vacuum and were difficult to make, but once they were available, new studies on electricity could be done. This led to the discovery by Julius Plücker (1801–1868) of cathode rays (named by Eugen Goldstein [1850–1930]). These rays would later become the foundation for television technology. Researchers found that the invisible rays could cause a glow, cast shadows when interrupted by a solid object, and be deflected by a magnetic field. William Crookes (1832–1919) demonstrated that the rays could exert a force and therefore were actually particles with mass. In apparent contradiction, experiments by Heinrich Hertz (1857–1894) showed that the cathode rays could pass through gold foil, which suggested that the rays were actually waves. In the 1890s, J. J. Thomson (1856–1940) did a series of studies in which he argued that what was zipping around the tube was a negatively charged particle that Thomson called a "corpuscle" in honor of Isaac Newton. Although Thomson objected to the name, the particle became known as an "electron."

The question for the elements was how the electron figured into the picture of the atom. Since matter could have a positive charge, a negative charge, or no charge at all, it was clear that the electron had to have something to do with matter, but was it a part of atoms (and thus part of elements) or was it something separate from atoms?

To make the problem even more perplexing, another aspect of physics seemed to be intruding into the study of matter. This was the problem of waves and particles. From the time of Newton, almost all matter theorists and most chemists had assumed that matter was made of solid particles. Light, according to Newton, was also a particle, but over time, experimental evidence mounted to suggest that light was a wave. For example, Thomas Young (1773–1829) did an experiment (now known as the double-slit experiment) around 1805 that demonstrated a beam of light could be split and show an interference pattern. By the beginning of the twentieth century, it seemed clear that light had a dual nature, acting like a particle in some cases (as when exerting a force) and as a wave in other cases (e.g., in having the ability to pass through itself). This strange situation was sometimes called the wave/particle duality. While this was a confusing situation, it seemed to be in the realm of physics and therefore not of any direct importance to the study of elements, which were clearly particles. Research into the electron and the electrical nature of the atom began to muddy the waters, however, since light and electricity were somehow related.

While the work on the electron was going on, another discovery came from the cathode tube. This was the 1895 discovery of X-rays by Wilhelm Röntgen (1845–1923). Röntgen had done his doctoral work on states of gas and was very interested in Crookes's tube, since it could be used to look at the behavior of gas at very low pressure and concentration. On November 8, 1895, while working with a Crookes tube completely covered with black cardboard, he noticed that a piece of paper coated with barium platinocyanide began to glow. Over the next two years, Röntgen explored and published on the properties of what he called X-rays, including their ability to pass through flesh. The first X-ray to show bones was made of the hand of Röntgen's wife, Anna Bertha, in 1886. For his discover of X-rays, Röntgen was awarded the first Nobel Prize in physics in 1901.

The discovery of X-rays did nothing to resolve the wave/particle argument that continued in chemistry and physics because X-rays clearly radiated (a wave property), but they also were generated by bombardment and seemed to have penetrating power (which made them seem more like a particle). In addition to the X-ray's potential as a medical diagnostic tool, it became a tool for the investigation of matter, especially in the development of crystallography. Scientists would eventual learn that by beaming X-rays at crystals, they could determine the arrangement of the atoms inside the substance. One of the greatest triumphs of X-ray crystallography was the determination of the structure of DNA in the early 1950s.

As a research tool, X-rays also advanced the study of the elements. Moseley's work from 1913 on X-ray spectra had shown that the order of the elements in the periodic table was the result not simply of chance but of some fundamental principle of atomic structure. It confirmed the ordering of some problematic elements and revealed that there was a missing element in the rare-earth series between neodymium and samarium. Element 61, promethium, was officially added to the periodic table in 1949.

Along with X-rays and cathode rays, the discovery of radioactivity would set the physical and chemical communities on new paths of research. Antoine Henri Becquerel (1852–1908) was doing research on the phenomena of phosphorescence and fluorescence, or the emission of light by materials. When he learned about X-rays, he wondered whether fluorescent materials could produce X-rays or cathode rays. In 1896, he discovered that certain uranium salts could darken a photographic plate through a lightproof wrapper and that the salts did not need to be exposed to light to fluoresce. One of the properties of the radiation from uranium salts was that it ionized the air and so could be detected as an electrical charge.

In addition to his scientific discoveries, Becquerel may have been the first person to suffer radiation burns. In 1890, he carried a vial containing a small sample of radium in the breast pocket of his jacket. He developed a wound that looked like a burn but that did not heal for several months. The danger of radiation to cells would not be understood for many years and was probably a factor in the death by cancer of a number of scientists.

Becquerel's student Marie Sklodowska Curie (1867–1934) and her husband, Pierre Curie (1859–1906), began to investigate this new phenomenon and were aided in their examination of the elements for radioactivity by the loan of some rare elements such as thorium. Only uranium and thorium seemed to have this property, but this led them to investigate the components of the mineral pitchblende. Pitchblende was composed primarily of uranium oxide, but the results of the ionization tests seemed to indicate that it had was more ionizing power than could be accounted for by uranium alone. After months of purification and separation work, in July 1898 they announced the discovery of polonium, named after Poland, Marie Curie's homeland. By December, further work led to the discovery of a second new element, which they named "radium." The Curies, along with Becquerel, were awarded the Nobel Prize for physics in 1903. Marie Curie also won the Nobel Prize for chemistry in 1911, remaining the only double winner until 1963, when the chemist and peace activist Linus Pauling won his second.

The discovery of radioactivity had a great impact on the conception of the elements. In addition to leading to the discovery of two new elements, radioactivity began to suggest that the elements were more dynamic than had been previously thought. There seemed to be some relationship between the physical makeup of matter and energy, since the radioactive elements, particularly radium, produced their own light, and there was also a great amount of electrical activity associated with the radioactive elements. In other words, radioactivity seemed to link energy to matter in a new way.

At about the time that the chemistry and physics worlds were being revolutionized by the discovery of radioactivity, the chemist Jean Perrin (1870–1942) was turning his attention to the question of the reality of atoms and molecules. He had already done work on the charge of cathode rays, so he was well aware of the complex problems the new discoveries had created. Perrin began a painstaking series of experiments, first studying the speed of sedimentation of tiny particles and then using the newly created ultramicroscope to trace the path of particles floating in a fluid. The motion of the particles was called Brownian motion, after Robert Brown (1773–1858), a botanist who first noted the motion of grains of pollen suspended in water.

The ultramicroscope, invented in 1903 by Richard Zsigmondy (1865–1929) and Heinrich F. W. Siedentopf (1872–1940), allowed scientists to look at particles at the very limit of optical observation. It used the Tyndall effect created by shining a very bright arc light set at right angles to the viewing lens. The particles were suspended in a liquid and observed against a black background. The light reflected from the particle in much the same way as dust motes can be seen in a ray of sunlight.

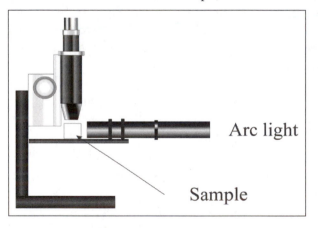

Arc light

Sample

Figure 18: Ultramicroscope.

What Perrin did was to create a model of a gas using tiny particles suspended in very pure water, reasoning that the particles he was watching followed the same kinetic rules as molecules in a gas. This all depended on atoms and molecules having a true material existence and obeying the laws of Newtonian physics. Since he created the particles, he knew their mass and dimensions, and by tracing the zigzag paths that they traveled, he could determine the force necessary to cause a change in direction. With that information he could calculate the physical characteristics of the gas molecules and also experimentally confirm Avogadro's number, which represented the number of molecules in one cubic centimeter of gas at standard temperature and pressure (STP).[1] Perrin's work, published in 1909, was a powerful demonstration of the physical reality of atoms and molecules. It did not, however, convince everyone, even if his results were difficult to explain using the theory of forces favored by the Energeticists.

What Perrin did not know at the time he was doing his work was that Albert Einstein (1879–1955) had also been thinking about Brownian motions. In 1905, Einstein published a paper outlining the conditions that would cause small particles to move in a diffused suspension and showing that if you could trace the motion of the particles, you could infer the number of molecules in a specific volume. In other words, Einstein's 1905 theoretical work and Perrin's 1909 experimental work fit together so well that the scientific world generally accepted the existence of atoms and molecules as actual physical entities. If atoms were real, then the unique characteristics of the elements were preserved.

With the discovery of the electron, radioactivity, and the proof that atoms were real, a new model of the atom was necessary. The Daltonian atom model, with its property of indivisibility, was no longer sufficient. The new model had to account for positive and negative charges, while still fitting with the characteristics that made up the periodic table, particularly atomic weight. Several models were proposed such as clusters or "dynamids," as well as a model called a "saturnian" atom because it had a positive inner sphere and a band of electrons. Another model was the "plum pudding" or "raisin muffin" that postulated a solid positive mass with negatively charged electrons (the raisins) imbedded in it.

Clarifying the structure of the atom required some method of getting information from the infinitesimally small object. For this, radioactivity went from being the object of research to a tool of research. In 1906, Rutherford noted that when a beam of alpha particles was aimed at a metal foil, the particles passed through the foil but showed a scatter pattern on a photographic plate. In 1908, Rutherford, working with his assistant Hans Geiger (1882–1945), the inventor of the Geiger counter, observed that a small number of particles that were beamed at the foil were deflected or even bounced back from the foil, while most passed through without effect. Geiger and Ernest Marsden continued the work, and in 1911 Rutherford described an atom that consisted of a central hard core (named the "nucleus" in 1912) that was positively charged and contained almost all the mass of the atom. The electrons orbited the nucleus at some distance, leaving the atom mostly empty space.

While Rutherford's model of the atom seemed well established in an experimental and theoretical sense, it also had some problems. The biggest problem was the electron death spiral. If the laws of physics applied to the electrons that orbited the nucleus, they should lose energy with every orbit and spiral down and crash into the nucleus. If this happened, the universe as we know it would not exist, so there had to be a problem with the model or with physics.

The great physicist Niels Bohr (1885–1962) had worked with Rutherford and decided to apply the new field of quantum theory to the model of the atom. Quantum physics had been introduced by Max Planck (1858–1947) in 1900 as part of an effort to resolve some odd results that physicists discovered with the way energy radiated. As it turned out, it was physics that had to change, not the model of the atom. The electrons did not lose energy while orbiting, so, although the Bohr-Rutherford atom (which remains the basic picture of the atom we use today) might look like a kind of miniature solar system, it did not act like one. The electrons had to be at specific distances from the nucleus, and, because of the energy level at each distance, there was a limit to the number of electrons that could be in each orbit, or shell. When an electron moved from one shell to another, it gained or gave up energy. That energy was not variable but came in set units. These packets of energy, known as quanta, were a single unit of energy or whole-number multiples of that single unit. While this solved a number of problems for the structure of the atom, it caused many headaches for physicists.

Electrons were easily separated from the atom, and scientists believed they understood at least the basics of how the electron fit into the model of the atom. After 1918 (much basic research on the atom was halted because of World War I), Rutherford noticed that some of the particles that appeared when an element was bombarded with alpha particles were heavier and positively charged. When nitrogen was bombarded with alpha particles, occasionally a strange particle appeared that seemed to have the same properties as a hydrogen nucleus. This suggested that it was coming from the nucleus (which had the vast majority of the mass of the atom) of the nitrogen and that the nucleus was, in effect, made up of hydrogen nuclei stuck together. Since hydrogen had already been established as have the atomic number one, Rutherford called the strange particle a "proton," from the Greek *protos*, meaning "first" or "original."

Thus, the most basic atom, and the most basic element, was made up of one electron (with a negative charge) and one proton (with a positive charge). The model seemed balanced and elegant, and it made Mendeleev's lighter-than-hydrogen element coronium impossible.

In terms of the elements, the new picture of the structure of the atom helped to explain why some elements, such as oxygen and carbon, were highly reactive, while others, such as gold or neon, did not combine with other elements, or did so only under extreme conditions. The reason was found in the number and arrangement of electrons. Each atom had a natural tendency to try to fill the electron shells to their capacity. Thus, the outer shell of carbon had space for four more electrons and could thus combine with four hydrogen atoms that had one electron each and were missing one electron to fill the inner shell. These bonds were called "covalent," a term introduced by Irving Langmuir (1881–1957) in 1919 to indicate the sharing of electron pairs. These ideas were presented to the world by Gilbert Newton Lewis (1875–1946) in his book *Valence and the Structure of Atoms and Molecules* (1923).

In 1920, Rutherford had speculated about the existence of another type of particle in the nucleus of atoms. This was based on the observation that the atomic weight of elements did not rise as a simple additive corresponding to the number of electrons. For example, hydrogen had one electron (weighing almost nothing) and one proton and an atomic weight of one. Helium had two electrons and two protons, but its weight was not two, but four. It was not until 1930 that experimental evidence was found to support the addition of another component to the structure of the atom. James Chadwick (1891–1974), in a complex experiment, detected the neutron. Chadwick, who had worked with Rutherford on the discovery of the proton, noted that some experimental results from the bombardment of light elements like lithium and boron could be explained only by the presence of a new particle. This was named the neutron, since it had no electrical charge.

The discovery of the neutron solved two problems that had confused scientists about the relationship between atomic weight and atomic structure. In addition to explaining why the masses of the elements did not increase by one unit for each step up the periodic table, neutrons also explained why experimental measurements of the masses of the elements did match the theoretical values predicted by the periodic table. Careful studies of the elements using mass spectroscopy showed that their masses were not whole numbers. The variation was often very small and was at first assumed to be contamination. As chemical techniques for analysis samples improved, however, it became clear that any sample of an element contained a certain portion of slightly different atoms. These were still of the same element, sharing chemical properties, but not quite the same substance, being a slightly different mass. These odd materials were called isotopes. In 1920, Francis William Aston (1877–1945) established that the average mass of the hydrogen atom was 1.008. With the addition of the neutron to the structure of the atom, isotopes were shown to be chemically similar atoms of an element with differing masses. This came from the fact that the atom had the same number of protons but a different number of neutrons. The slight difference in mass would be an important factor for the development of nuclear power and the atomic bomb. It was also found that some isotopes were more stable than others.

While Lewis's and Longmuir's ideas were eventually superseded as physicists looked more deeply into the parts of the atom, the basic model of chemical activity largely remained based on the Bohr-Rutherford atom using electrons, protons, and neutrons. The periodic table was no longer just an ordering of the elements by characteristics but a representation of the physical structure of the atom itself. Chemical activity, atomic weight, the existence of isotopes, and the periodic nature of the distribution of the elements were neatly explained. There are a number of naturally occurring isotopes of many elements, particularly at the heavier end of the periodic table. The most famous isotope is uranium-235 (often written as ^{235}U), the material used to make the first nuclear bomb.

At some level, it seems that the story of the elements should have been completed by 1930, but it was not. There was another aspect to the discovery of radioactivity that made research on the elements not just possible but a hot topic to the present. The principle question about radioactivity was why radioactive materials emitted the energy and particles that they did. When Rutherford and Geiger used alpha particles to examine the structure of the atom, they were using a small amount of radium or radon gas that emitted alpha particles. The particles had to be coming from the atoms, so, in effect, the atom was changing from one substance into another by itself. Rutherford compared the decay of radioactive material to a porcelain figurine tumbling down a set of stairs. As a figure falls, bits and pieces come off until it comes to rest at the bottom of the stairs. Similarly, extremely heavy radioactive elements were slowly turning themselves into lighter, more stable elements. Radium, for example, closely resembles lead. What decay suggested was that, over time, radium would actually become lead. Rutherford would present a number of these ideas about transmutation to the public in a book called *The Newer Alchemy* (1936). He chose the title to reflect that fact that his model of the physics and structure of the atom seemed to make the transmutation of matter not just possible but a natural and constant event in the real world.

The introduction of time as a necessary consideration for understanding matter was a radical change to the conception of the elements, which since ancient times were thought to be eternal. To describe the life of a decaying radioactive element, in 1900 Rutherford introduced the concept of the "half-life." The half-life of a substance is the time it takes for half of the atoms in a sample to decay (emit nuclear radiation) and, in effect, for the sample to become a different substance. Half-life can also be used to express the time it takes for the chemical or physical activity of the sample to decrease by half.

Since the heavy elements tended to have several isotopes, each with a different half-life, finding new elements depended on collecting isotopes that had half lives long enough to be tested. The time of the half-life was also a way to identify materials. Half-life calculations could be used to identify a substance and to estimate how long it would exist, but it could also be used in the other direction: By measuring how much of a particular isotope was found in a sample, you could calculate how long ago it had been something different and decayed into the present form. For example, uranium-238 decays into lead-206. By measuring the amount of lead-206 in a rock, you can calculate when the rock was formed. So if you found 100 lead-206 atoms in a sample and no uranium, you would know that at some distant time there had been 100 uranium-238 atoms and no lead. If you found 70 lead-206 atoms and 30 uranium-238 atoms, you would know that less time had passed, since not all the decay had taken place. Since the rate of decay is constant, you could make a good estimate of how much time had passed.

In archeology and forensics, carbon-14 (half-life of 5,700 years) dating has been used extensively, since living things absorb carbon, but when they die, no new carbon is absorbed and the rate of decay of the absorbed carbon gives a good measure to the time of death. This can be used only for a certain length of time (about 60,000 years), but other isotopes, such as potassium-40 (half-life around 1.3 billion years) and thorium-232 (half-life about 14 billion years), extend the range considerably.

The concept of the half-life also led to more problems in physics, since the rate of decay was constant but the time for any particular atom to emit a particle and change could not be known for certain, only presented as a probability over time. The idea that the universe could be known only through probability struck many scientists as bizarre and against the very idea of science and its search for certainty.

The most famous example of this problem was proposed by Erwin Schrödinger (1887–1961) in 1935, when he wrote about a "diabolical device." Schrödinger suggested the following thought experiment. Place a cat in a sealed and opaque box. Inside the box is a small sample of a radioactive material that in the course of one hour has a 50/50 chance of decaying. If there is decay, a sensor will release a poison gas that kills the cat. The problem is not with the concept of probability or radioactive decay but with the state of the cat at the end of the hour. Is the cat alive or dead? The only way of determining that is to look in the box, but, until the lid is opened, the cat is, from the point of view of physics, equally alive and dead at the same time. In other words, the state of the universe might be completely indeterminate until someone looks at it.

Today, the problem of certainty versus probability still exists and remains a philosophical problem, although, as it turned out, it has in no way stopped research. Scientists, especially physicists, were forced to change the way they expressed some ideas, but in some ways it made science more accurate, since it became easier to describe and explain the variability of nature.

A second aspect of the use of radioactivity was that elements could be changed by being bombarded with various particles. In 1932, a team working at Cambridge University bombarded lithium with hydrogen nuclei (protons) and produced helium. These kinds of processes were also used to create artificial radioactivity by bombarding a nonradioactive element and converting it into an unstable isotope of a different element. The unstable isotope would then emit particles and become yet another element. Marie Curie's daughter, Irene Joliot-Curie (1897–1956), and son-in-law, Jean-Frédéric Joliot (1900–1958), transmuted aluminum into phosphorous, and then the phosphorous decayed into silicon. In a sense, the quest of the alchemist seemed to have been achieved. Chemists and physicists could produce controlled transmutation of the elements. The modern alchemy revealed not the secret of the Philosopher's Stone but rather the natural process by which matter was built up and broken down. While the true alchemists would have been pleased with the discovery, those hoping to turn base metals into gold would have been disappointed—the cost of such a transmutation would far outweigh the value of the few atoms of gold that might be made.

Bombardment experiments were being carried out in a number of laboratories around the world, most notably in Italy, Germany, France, England, and the United States. There was something of a race to build bigger and more powerful equipment to do the bombarding, as well as a race to see what exotic products could be generated. Irradiation by exposure to a source of particles began to be replaced by irradiation by a variety of machines, such as linear accelerators and cyclotrons. These devices could accelerate particles, particularly protons and neutrons, and send a beam of concentrated particles at a small sample of an element (or any other substance). The beams could do a number of things, such as knock parts of the target off or force beam particles into the nucleus. As the machines became more powerful, the beams were also able to break apart the subatomic particles to reveal even smaller particles such as muons and gluons.

The basic principle of the accelerator is simple. A particle is sent into a tube, where it passes through a series of fields of force. These fields can come from different parts of the electromagnetic spectrum, so that electrical, magnetic, radio, or a combination of fields can be used. As the particle passes through the field, it gets a little push (imagine pushing someone on a swing or a carousel). With each push, the particle goes faster and faster, gaining energy. This energy, measured in electron volts and, at higher levels, in MeV, or mega electron volts, and even GeV, or giga electron volts, is necessary to overcome the forces that hold the parts of the atom together and, at the highest levels, to overcome the forces that hold the parts of the particles together. The collisions produce small amounts of new material as well as radiation traces that can be recorded and used to determine what the products are and to provide information about the interior of the atom.

The first device to accelerate electrons was created in 1928 by Gustaf Ising (1883–1960). It passed particles through electrical fields. During the 1930s, many laboratories used Robert Van de Graaff's (1901–1967) electrostatic generator, which produced up to 10 million electron volts. Ernest Lawrence (1901–1958) created the cyclotron, which used two large magnets (called "dees" because of their shape) to propel particles in a circular orbit as they accelerated. It had the advantage of not needing super-high voltage to work, although it still drew large amounts of power. Lawrence's first large cyclotron, built with 27-inch magnets, was completed in 1931. Work on new types of particle accelerators stopped during the war years, but during the Cold War there was a kind of accelerator race as different nations strove to build bigger machines. In 1946, the Cosmotron was completed, which could accelerate protons with 2.5 GeV. In 1955, the Soviet Union announced the development of a 10 GeV synchrotron, to which the United States responded with plans for a 12 GeV synchrotron. A number of European nations formed the European Centre for Nuclear Research (better known as CERN) and announced that they would build a 25 GeV synchrotron. By the 1980s, the power of these devices had reached 400 GeV, and there were plans to build a massive superconducting supercollider at a cost of $11 billion, but the project was eventually canceled as being too expensive.

One of the leading research teams looking into the products of bombardment was composed of Otto Hahn (1879–1968), Lise Meitner (1878–1968), and Fritz Strassmann (1902–1980), working in Berlin. They irradiated uranium with neutrons for 80 days and found radium, barium, lanthanum, and zirconium. The radium then seemed to decay into actinium and the actinium into thorium. In 1938, Meitner was forced to flee Germany when Germany annexed Austria and her legal status as a Jew came under German law. Hahn and Strassmann continued their work on the bombardment of uranium and noticed some odd results, producing, in particular, barium. The researchers suspected that they had split the nucleus of the uranium, but they needed other evidence. Barium has an atomic weight of 140, slightly more than half the atomic weight of uranium, so Hahn and Strassmann looked for by-products in the atomic weight range of 80 to 100. They found strontium and, later, krypton and rubidium. Because no one else had thought to look for these by-products, no one else had conducted the tests necessary to find them.

Hahn and Strassmann reported on their finding to Meitner. In discussion with Otto Frisch (1904-), her nephew and a physicist, Meitner recognized that the neutron bombardment had split the uranium nucleus. She called this process "fission," and, in January 1939 with Frisch, she sent a letter to the journal *Nature* announcing the discovery of nuclear fission. The splitting of the atom was accompanied by the release of neutrons and the conversion of a small amount of mass to energy. The amount of energy could be calculated using Einstein's famous 1905 equation $E = mc^2$, and a new source of power was born.

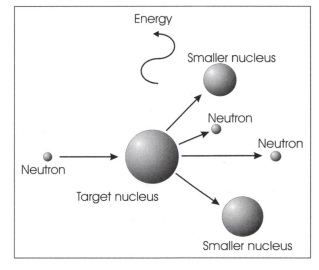

Figure 19: Fission.

The discovery of fission did not, by itself, lead to the atomic bomb. Rather, it had to be combined with another idea, one that came from the physicist Leo Szilard (1898–1964). Szilard was also a refugee from Fascist Europe, and, according to one story, while walking in London one day, he came up with the idea of the chain reaction. If a neutron hit the nucleus of a heavy atom and caused the atom to release more neutrons, there could be a growing and never-ending cascade. Combined with the information that nuclear fission had been achieved, this idea made an atomic weapon a possibility. Szilard, Einstein, and other scientists lobbied the American government to begin a project to build a nuclear weapon, fearing that the German scientific establishment, although weakened by the loss of many refugees from the Nazi regime, was still powerful enough to build such a weapon. The threat of a German bomb came one step closer when Belgium surrendered to Germany, placing the Belgian Congo in the hands of Hitler; in 1940, the Congo was the greatest source of uranium in the world. Concern about the power of Germany led to the creation of the Manhattan Project.

One of the biggest jobs for the Manhattan Project was gathering enough of the right kind of uranium to make a bomb. The development of the first nuclear weapon depended on the collection of the isotope ^{235}U from uranium ore that was primarily made up of ^{238}U. Although ^{238}U was radioactive, ^{235}U was far more fissionable, which was critical to create the chain reaction. The separation process was difficult and time consuming, since ^{235}U isotopes represent only about 0.7 percent of the atoms in any given sample of uranium. Because they were the same element, the uranium atoms could not be separated by chemical means. The only difference was in mass. Several methods were tried, but, in the end, most of the ^{235}U was collected using gaseous diffusion. Uranium hexafluoride gas was squeezed through clay filters. The lighter isotope would tend to go through first, so it was collected and filtered again and again until only the desired isotope was left.

The use of bombardment and the radioactive decay of uranium allowed the last of the empty spaces in the periodic table to be filled. These were elements 61, 85, and 87. In 1939, Marguerite Perey (1909–1975), working at the Institut du Radium, in Paris, traced the decay of ^{235}U and found isotope $^{223}87$. Although the element decayed quickly, it lasted more than 20 minutes, enough time to test its properties and to confirm that it was the missing element. In 1946, Perey named the element "francium." Element 85 could be created only by using a cyclotron; an American team in California bombarded bismuth with alpha particles and isolated an isotope that had a half-life of around seven hours. The element was named astatine, from astatos, meaning "unstable."

Promethium was the last of the suburanium elements to be added to the periodic table. From as early as 1902, there had been speculation that there should be an element between neodymium and samarium, in spot 61. Although there was one claim to have found element 61, in 1924, and a second claim by a different group, in 1926, these claims could not be verified (and likely could not be authentic, given the processes used). It was not until 1938 that L. L. Quill, working with a cyclotron at Ohio State University, produced a small amount of an isotope that fit most of the characteristics for element 61, which they proposed be called "cyclonium." Because of problems with Quill's results, the official credit for creating element 61 went to L. E. Glendenin and J. A. Marinsky. They proposed the name "prometheum," which was adopted with a slight change in spelling to "promethium," in 1949.

During World War II, more work was done on the elements, particularly as part of the Manhattan Project. After the war, the existence of these new elements was made public. All of the naturally occurring elements from hydrogen to uranium, including the small number of artificially created but low-number elements francium, astatine, and promethium, had been found and put into the periodic table. Combined with the work of Rutherford and Bohr, this work made the physical structure of the atom clear and proved beyond all doubt. Even if physics would start to find smaller parts within the electron, the neutron, and the proton, from a chemical point of view, the atom was complete. Atoms and molecules had been shown to be real, they way they combined was understood, and the energy inside the atom was not only known but would be released, first in an atomic pile (what we would call a nuclear reactor) and then, in 1945, as nuclear weapons. The concept of the element had been at the heart of the formal study of matter for more than two thousand years and was one of the great driving forces in science. The hunt for the elements had affected the theory of matter, led in part to the transition from alchemy to chemistry, been a key component in the development of experimental method, and opened the door to understanding not just individual elements and their characteristics but matter itself.

Despite the success of the chemists and physicists in unraveling the secrets of the elements, this was not quite the end of the story. The next step in the story of the elements takes us out of the realm of chemistry and into the realm of high-energy physics. The question of what the elements were had been solved, at least at the practical level of the material universe, but it opened a door to a related question: How did the elements that made up everything come into being? Why were there elements at all? The search for the answers to those questions would keep the study of the elements a hot topic in science.

NOTE

1. STP or standard temperature and pressure is one atmosphere at 0° Celsius.

10

THE ELEMENTS BEYOND 92

The concept of the elements depended on two different but ultimately complementary ideas about matter. The first idea was ancient: that the elements were the fundamental building blocks of nature. Whether there were 1, 2, 3, 4, or 92 elements was in a sense less important than the power of the concept to explain nature and direct research. The second idea came with the discovery of the structure of the atom and the physics that made that discovery possible: that an element represented a specific combination of subatomic particles determined by physical laws. The creation of controlled nuclear fission and the invention of accelerators and cyclotrons made a kind of modern alchemy possible, allowing the creation of new elements that were not found in nature but that still met the new conditions to be considered elements.

In 1940, element 93 was discovered, or created, depending on how experimental research is described. The Berkeley team of Edwin McMillan (1907–1991) and Philip Abelson (1913–2004) bombarded uranium foil with neutrons and found that one of the fission products lasted for 2.3 days before naturally decaying. This did not match with known elements, and they suspected it was a new element. With 2.3 days to conduct tests, they were able to test oxidation states and positively identify that it was unique. The name "neptunium" was suggested because the planet Neptune was beyond Uranus.

Glenn T. Seaborg (1912–1999), one of the leading scientists of the Manhattan Project, was a great element hunter. Seaborg studied neptunium, creating 45 micrograms of a more stable isotope using a cyclotron and samples created in a nuclear pile. By creating neptunium from uranium, he then formed a new element. This was element 94, and the isotope created was calculated to have a half-life of around 24,000 years, making it a very stable element compared to some of the other created elements. It was named "plutonium" after the planet next farthest out. This work, and the creation of the elements americium (95) and curium (96), was kept secret until the end of World War. II

The creation of plutonium was significant beyond its scientific importance. Plutonium was highly fissionable, and it was used by the Manhattan Project as the fuel source for the second bomb (Little Boy) dropped on Japan at the end of the war. Getting enough plutonium to make a bomb proved to be difficult. The cyclotron method was far too slow, and so the plutonium was mostly generated using fission in an atomic pile. Nuclear reactors that are set up to produce plutonium and other radioactive isotopes are called "breeder reactors." The ability to create weapons-grade plutonium and other fissionable materials continues to be one of the sources of fear about modern day of nuclear power. While all nuclear reactors can theoretically be used to transmute matter, some reactor designs are much better for transmutations. Since materials for use in cancer treatment, high-energy physics, advanced power systems, and other peaceful uses can be made by these reactors, reactors that are better for transmutation are very desirable. It is difficult, if not impossible, to separate the peaceful uses from the military uses.

Seaborg's experience synthesizing elements led him to propose the creation of a separate section or subgroup of the periodic table. The lanthanide series had already been established and was considered to be almost a mini periodic table, covering the rare-earth elements from lanthanum (57) to lutetium (71). It had been created to fit the rare-earth elements into the periodic table without knocking the rest of the table out of order. By 1944, Seaborg was proposing another series, called the actinide series, that would cover actium (89) to element 103. Theoretically, element 104 would fit back into the characteristics of Group IV, with titanium (22), zirconium (40), and hafnium (72). Following the rules of the periodic table, elements 104 to 117 would have characteristics like the other natural elements in their groups, with element 118 at the end of the new period (row) being a noble gas like argon (18) or xenon (54). A new subgroup, the super-actinide series, would appear at element 122.

The hunt for new elements can happen only at the high end of the periodic table, and the idea that there could be superheavy elements (SHEs) started to be discussed around 1955 by John Wheeler and Gertrude Scharff-Goldhaber. Theoretically, new elements could be created indefinitely as neutrons, protons, and electrons are brought together in greater numbers. While the additions are not simply additive, the specific rules of the atom allow scientists to calculate the structure of any theoretical element. However, as the new synthetic elements went further up the table in terms of mass, the energy required to create them also increased. Although larger and larger cyclotrons were being constructed, in part to look for new elements, the next three new elements were found not in the laboratory but in the residue of test blasts of thermonuclear weapons. The elements einsteinium (99), fermium (100), and mendelevium (101) were created in the fiery explosion that copied the energy production of the sun.

The physics of fusion was the opposite of fission. Rather than breaking atoms apart, the great energy and pressure of a fission explosion triggered the fusing of hydrogen nuclei. This was the same process that fueled the power of the sun and stars.

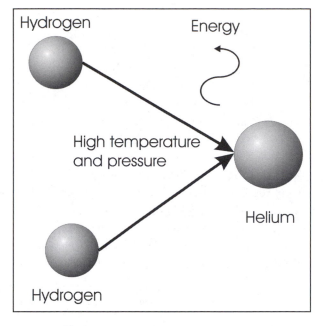

Figure 20: Fusion.

One of most important aspects of the Cold War was the race to develop and deploy nuclear weapons. The early arms race was largely focused on nuclear fission, but by the 1950s the much more powerful fusion weapons were the main area of development. The competition, however, was not only military, and the quest for new elements became a minor but important competition, primarily between American laboratories and Soviet ones. Because the equipment used to do nuclear research was used to produce elements and material for weapons, the link between the synthesis of new elements and the arms race was direct: if one side could produce a new element, it revealed that that side's equipment, resources, and scientists had an advantage over the opposition. Since the equipment, resources, and scientists were also part of the nuclear arms development system, the implication was that the advantage would extend to the weapons. Even without the arms race, the competition to synthesize new elements was fierce, since international scientific status and even Nobel Prizes could be gained by such work.

Between 1958 and the mid-1970s, every announcement of a new element was contested. Problems with priority of synthesis, proper identification, theoretical errors, disputed testing methods, and alternative names plagued each addition to the periodic table. When, in 1958, the discovery of the element nobelium (102) was announced by Ghiorso and Seaborg, the Soviet scientists Georgli Flerov and E. D. Donets and others cast doubt on the claim. Again, in 1961, when the discovery of lawrencium (103) was announced, there was criticism from Soviet scientists. When Soviet scientists presented evidence for the synthesis of element 104, in 1966, and named it "kurchatovium," American objections blocked its acceptance and then prepared their own synthesis of element 104, calling it "rutherfordium" in 1969. Another fight over element 105 lasted until 1970, with the Soviet Union using the name "nielsbohrium" and the Americans proposing "hahnium."

In order to avoid the growing tension and confusion caused by naming rights, the Union of Pure and Applied Chemistry (IUPAC), an international body made up of representatives from almost every country in the world, declared that no new names for elements would be officially recognized after lawrencium (103). Instead, they set up a naming system based on a numerical code plus the ending "ium."

Following this system, element 108 would be "uniloctium" (dropping the redundant "n" from the 0/nil part) and the symbol "uno," while element 130 would be "untrinilium" (utn) and element 201 would be called "binilunium" (bnu). Recognizing the historical role of names, IUPAC also agreed that the discoverers of new elements could present a "trivial" common name that would not be used in official scientific references to the new elements. Although some periodic tables follow the IUPAC principles, in reality, many tables continue to use the "trivial" name for elements above 103, and naming continues to be one of the rewards of element discovery.

Element 106, named "seaborgium" to honor Glen Seaborg, was created in 1974 by the Flerov Laboratory of Nuclear Reactions in Dubna, in the Soviet Union, and at the Lawrence Berkeley and Livermore Laboratories, in the United States. Seaborgium isotopes are created using a cyclotron and can exist from about 0.9 seconds up to about 20 seconds before they decay into other elements. This is enough time to determine the properties of seaborgium and to confirm that it qualifies as an element.

The two most recent new elements to be recognized are elements 110 and 111. Element 110, named "darmstadtium" and symbol Ds, was first identified in 1994 and officially recognized in 2003. It was named for the city Darmstadt, in Germany, where it was created. To make darmstadtium, scientists fused an atom of nickel with an atom of lead. The resulting atom lasted for only

Number	Name
0	nil
1	un
2	bi
3	tri
4	quad
5	pent
6	hex
7	sept
8	oct
9	enn

1/1000th of a second, but by studying the building process and the decay products, the original superheavy element was verified. Element 111 (ununu-nium) was officially recognized in 2003, although the suggestion to name the element "roentgenium" with the symbol Rg has not yet been approved.

It was predicted by various scientists, including Wheeler and Scharff-Goldhaber, that new elements above uranium would have shorter and shorter half lives, breaking down into other elements by nuclear decay, and this was confirmed by the experiments on the synthetic elements during and after the war. The theory of Super Heavy Elements also suggested that there might exist islands of stability where the new elements would be stable for longer periods of time, perhaps even long enough to make them more than objects of scientific research. There would be two stable points at element 114 and 126, and perhaps a whole range of stable elements between 122 and 153. The hunt for superheavy elements began in the 1960s, but the limitations of available equipment restricted the work that could be done. Essentially, it takes huge amounts of energy to force particles together in the nucleus of an atom. As the number of particles grow, so does the complexity of their arrangement and the repulsive forces involved. There is an analogy with magnets. It is relatively easy to force two south poles together when the magnets are small and weak, but trying to force a small bar magnet to stay connected to a giant junkyard electromagnet would take far more energy and could easily be beyond the strength of a human being.

As of the writing of this book, claims have been put forward for the creation of elements 112, 114, 116, and 118, but none of these claims have been completely validated and the 2001 claim for the discovery of 118 was later retracted by the research team. IUPAC and IUPAP (the International Union of Pure and Applied Physics) continue to oversee and adjudicate claims and in 2003 released a Technical Report saying that although many of the claims were based on good science and sound methods, there was not yet enough independent confirmation to establish priority. In the case of these elements, it is probably just a matter of time before they are officially recognized.

Even though linear accelerators and cyclotrons have increased in size and power and combinations of accelerators and cyclotrons have been built, even bigger devices will be needed to continue the hunt for elements. These machines cost millions of dollars to build and operate, and, rather than scientific limits, it may be financial limits that slow research. The controlled use of fusion could potentially overcome these problems. In nature, all the elements above hydrogen are the by-product of fusion in stars, either directly or indirectly by radioactive decay from heavier elements created by fusion. All life on Earth and the Earth itself are made up of elements that started as hydrogen atoms that were fused together at incredible pressure and temperature. The helium atoms produced were then fused together, and so on. We can even observe some aspects of this process of creation by looking at the spectral images of different stars. Each one has a unique spectral signature, depending on how much material in addition to hydrogen is contained in the star.

Scientists around the world have been working on controlled fusion for many years. The theoretical and practical problems are considerable, but if fusion could be made to happen, one of the potential products would be the ability to custom-make elements.

The scientific interest in the elements since the middle of the twentieth century has been closely linked to the development of physics and the demands of the Cold War arms race. As chemists and physicists looked into the interior of the atom, they learned more about the structure of matter, and that knowledge, in turn, allowed a degree of control over the creation of matter. Nuclear reactors can now produce both useful and deadly materials even as they put a new source of power in to the hands of people around the world. The "atom smashers," as the cyclotrons and accelerators were sometimes called, made new elements possible, and scientists continues to explore the elements, looking for the islands of stability among the superheavy elements.

CONCLUSION

The history of the elements is in many ways the history of chemistry and physics. Before the search for the elements, there was the philosophical question of what was an element. That question was posed more than 2,500 years ago, and it has had many solutions. Each in turn has given way to a new answer that was made possible by new theories, new tools, and new observations. Our current answer to what constitute an element looks like it has staying power, since it fits so well with so many other aspects of our understanding of chemistry and physics, but there are still aspects of matter that are unresolved. Future work on the origin of matter in the universe, work on dark matter and anti-matter, on superstrings and the interior of subatomic particles, could alter our understanding of the elements.

The search for the elements and the effort to define what an element actually is have shaped our understanding of the world around us and have shaped our world. In science fiction, amazing or previously unknown elements are often introduced to enhance the limitations of nature. The best known is the dilithium of *Star Trek,* a substance that makes faster-than-light, or warp-speed, travel possible. While these imaginary elements are more fiction than science, the search for and the discovery of the actual elements has changed human history. Everything from the manufacturing of iron weapons to the catalytic converter for the exhaust system of an automobile (using platinum as a catalyst), from the discovery of oxygen to radioisotopes of iodine used in nuclear medicine, has depended on finding elements, understanding their properties, and then putting the elements to use. Along the way, the structure of matter was revealed, and the basic rules of the material universe came to be understood. Although we largely take the periodic table for granted today, a closer look shows that it really is both a guided tour of science and a trip through history.

Looking at periodic tables from different times and produced in different countries reveals a long history of scientific and political struggle. Even today, periodic tables demonstrate the creator's position regarding the control of chemistry by a governing body through its decision whether to use the IUPAC system or the names declared by the discoverers. Despite the various controversies, the periodic table of elements has become a standardized tool for education and research. It can be found as giant wall charts in classrooms and on wallet-size laminated plastic charts, and there are interactive versions on the Internet. It seems that the question of what the elements are and how they work has been answered. As far as everyday life is concerned, the elements listed on the periodic table really do constitute the matter that makes up the world around us, at least to element 92.

The story of the elements is also a powerful illustration of the interaction of esoteric knowledge and utility. Philosophers and blacksmiths each contributed to the story. Alchemists wanted to gain the secrets of nature for the wonder of such knowledge, and they also were interested in profiting from their work. Many of the greatest thinkers in human history saw the study of the elements as a way to glorify God and to understand God's creation. The knowledge gained by studying the old and new elements has traveled far outside the laboratory and has made the high-technology world of today possible. Computers, MP3 players, bar-code scanners, and televisions are possible only because of the discovery and understanding of the elements. Equally, the treatment of cancer, the development of pharmacology, and the use of CAT scanners are a product of the knowledge gained by studying the elements. Using our knowledge of the elements has even given us knowledge of distant stars and galaxies and allowed us to explore the moon and Mars and to learn about the composition of our neighbors in the solar system.

The building of the road from Thales's three prime elements to our modern periodic table listing more than 110 elements represents one of the great achievements in human history.

APPENDIX 1

THE PERIODIC TABLE

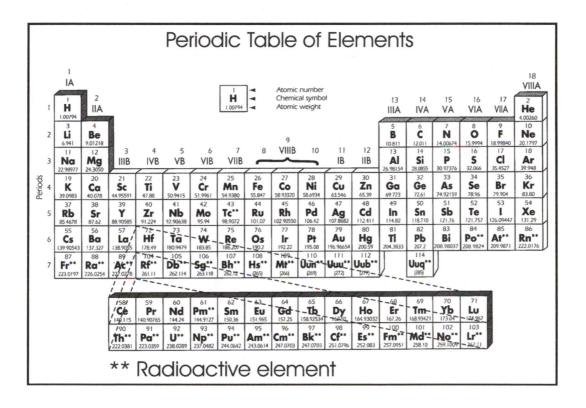

Periodic Table of Elements

** Radioactive element

HISTORY OF
THE ELEMENTS

The following gives a brief history of the discovery of the individual elements arranged chronologically, with descriptions of the origin of the name and symbol and common uses of the element.

Determining when an element was discovered can be complex, since the elements of the periodic table were not "elements" until the concept of an element was codified in the eighteenth century, particularly following the work of Antoine Lavoisier. Where there are multiple dates, they represent the first clear date of identification of a substance that would later be included in the periodic table, followed by a date representing the determination of the substance's isolation in a form we would consider elemental, thereby qualifying it as a member of the periodic table.

ELEMENTS KNOWN IN ANTIQUITY

Carbon

Symbol: C
Atomic number: 6
Atomic weight: 12.011
Melting point: 3550° C
Boiling point: 3800° C
Atomic radius: 77 pm

Carbon was used in various forms from very ancient times. The controlled production of carbon as charcoal may date from as early as 1 million years ago. The name comes from the Latin *carbo*, meaning charcoal. Carbon exists widely in nature and is known to be present in stars, comets, and the atmospheres of other planets. It is found frequently in combination with other elements, such as oxygen (carbon monoxide and carbon dioxide), as well as in the form of organic material, which by definition must contain a carbon atom. In the form of diamonds, carbon is the hardest naturally occurring material.

Gold   Symbol: Au
Atomic number: 79
Atomic weight: 196.96
Melting point: 1064° C
Boiling point: 2856° C
Atomic radius: 144 pm

Gold was one of the first elements to be mined and smelted. It has been prized for its luster, rarity, malleability, and resistance to tarnishing. These qualities make it an ideal material for jewelry and coinage. Its name in English comes from Anglo-Saxon, while its symbol comes from the Latin name *aurum*. Gold is found in few places in commercially exploitable quantities, and it has been estimated that all the world's gold would fit in a cube with sides of sixty feet. Its ability to conduct electricity has made it important in the modern electronics business.

Silver   Symbol: Ag
Atomic number: 47
Atomic weight: 107.9
Melting point: 961.8° C
Boiling point: 2162° C
Atomic radius: 144 pm

The mining and smelting of silver probably dates back to at least 3000 B.C.E. Like gold, it has properties that make it highly desirable for jewelry and coinage. Although silver tarnishes when exposed to ozone or air containing sulfur, its general chemical neutrality makes it very useful for science and commercial applications. Silver has the highest electrical and thermal conductivity of all metals. Its English name came from the Anglo-Saxon *seolfor*, while its symbol came from the Latin *argentum*.

Copper   Symbol: Cu
Atomic number: 29
Atomic weight: 63.5
Melting point: 1084.6° C
Boiling point: 2562° C
Atomic radius: 127.8 pm

Copper is the oldest mined and smelted metal. Traces of copper works date back to at least 5,000 B.C.E., and it was likely from the accidental discovery of copper smelting that the idea of smelting and alloying metals was developed. The name and symbol come from the Latin *cuprum,* in turn from the name of the island of Cyprus, a source of copper in the ancient world. There are large deposits of copper around the world. Its malleability makes it a useful metal, and it was used in jewelry and coinage throughout history. It was in combination with tin to produce bronze that copper changed history. The so-called Bronze Age, starting around 3000 B.C.E. was the first mass industrial exploitation of metal.

Sulfur   Symbol: S
Atomic number: 16
Atomic weight: 32.066
Melting point: 119.6° C
Boiling point: 444.60° C
Atomic radius: 104 pm

Sulfur in various forms has been used by humans for several thousand years. Its name comes from the Sanskrit *sulvere* and its Latin root, *sulpur*. Sulfur and sulfur compounds occur widely in nature, but sulfur is often closely associated with volcanic activity and is found in hot springs. Burning sulfur produces an unpleasant odor, and in high concentrations it can be toxic. Historically, sulfur was known as brimstone. Highly reactive, sulfur was an important chemical for alchemists, and Paracelsus based part of his system of the elements and medicine on sulfur. The industrial use of sulfur includes the production of gunpowder, fertilizer, rubber (vulcanized rubber), and fungicides, as well as food preservation. The largest modern use of sulfur is for the production of sulfuric acid, which has many commercial uses. Sulfur has become an environmental concern because it is released into the atmosphere when hydrocarbon fuels, particularly high-sulfur coal, are burned.

Tin   Symbol: Sn
Atomic number: 50
Atomic weight: 118.69
Melting point: 231.93° C
Boiling point: 2602° C
Atomic radius: 140.5 pm

The name tin comes from the Anglo-Saxon *tin*, while the symbol comes from the Latin name *stannum*. Like copper, tin was mined and smelted in antiquity from at least 3500 B.C.E., both for itself and as a component in bronze (80 percent copper/20 percent tin). Elemental tin is uncommon in nature, but the tin ore cassiterite (tin oxide) is found fairly widely. Because one form of tin is very malleable, it has been used for a wide range of domestic products, such as plates and cups, and in more recent times as tin cans and foil. Tin loses it malleability and can actually crumble to powder in temperatures below 13° C, making it a poor choice for many applications. This perishing of tin is called "tin disease." Tin has largely been replaced by other materials in domestic applications, but, as an alloy, tin continues to be used in bronze, solder, and pewter and as a glaze for ceramics.

Lead   Symbol: Pb
Atomic number: 82
Atomic weight: 207.2
Melting point: 327.46° C
Boiling point: 1749° C
Atomic radius:175 pm

Lead is another ancient metal. The name comes from the Anglo-Saxon *lead*, while the symbol comes from the Latin *plumbum*. Lead's extreme malleability and low melting point made it very useful, especially for casting objects such as figurines or, in more modern times, printing type. Elemental lead is rare in nature, but smeltable ore, primarily galena, can be found in a number of places in large quantities. Lead's high density meant that it was used as weight for things like ship ballast, while its resistance to many acids made it a useful material for industrial processes. White lead has been used since ancient times as a pigment, both in makeup and paint, but its use has been restricted or banned because of its toxicity, which can cause brain damage.

Mercury   Symbol: Hg
Atomic number: 80
Atomic weight: 200.59
Melting point: −38.83° C
Boiling point: 356.73° C
Atomic radius: 160 pm

Mercury has been found in Egyptian tombs dating back to 1500 B.C.E. Also known as quicksilver (using "quick" to mean alive), the Latin root came from the god *Mercurius*, the periodic symbol comes from the Greek words *hydoor*, meaning living, and *argyros*, meaning silver. Mercury is the only metal that is a liquid at room temperature, and it rarely occurs in elemental form in nature. Most production of mercury comes from processing the mineral cinnabar. In the past, mercury was often used as a medicine, particularly in ancient Indian medicine and by Paracelsus in Europe, but in modern times it has been recognized as a dangerous toxin and environmental pollutant, especially in the form of methyl mercury. Mercury has many industrial applications, including thermometers, mercury-vapor lamps, mercury switches, explosives, and dental compound.

Iron   Symbol: Fe
Atomic number: 26
Atomic weight: 55.847
Melting point: 1538° C
Boiling point: 2861° C
Atomic radius: 124.1 pm

The name iron comes from Middle English, while the symbol comes from the Latin *ferrum*. Smelting of small quantities of iron date back to around 1200 B.C.E. (the beginning of the Iron Age), and, after the technique was developed, iron replaced bronze as the primary metal for tools and weapons. The use of iron weapons has long been cited as the reason for the power of the Hittite empire, although it is now clear that other factors were more important than just the secret of iron production. Iron is abundant in the universe, and it is found in siderites (a type of meteorite), which were known as "lodestones" and were the first magnets discovered by people. The core of the Earth is thought to be composed primarily of iron. Today, iron is used in a huge range of products, and, in fact, more than 90 percent of all metal refining is of iron. A large portion of iron production is for steel, which combines iron with coke (a form of carbon) and other trace elements, such as nickel and tungsten.

DATED DISCOVERIES

Arsenic 1250/1649
Symbol: As
Atomic number: 33
Atomic weight: 74.9216
Melting point: 817° C
Boiling point: 603° C (sublimation)
Atomic radius: 125 pm

Arsenic compounds were known in antiquity, and the name is derived from the Greek *arsenikon* and the Latin *arsenicum*. Elemental arsenic may have been discovered by Albertus Magnus in 1250, but this is unclear. In was Johann Schroeder in 1649 who published two methods for producing arsenic, after isolating elemental arsenic. Arsenic was historically used in medicines and was developed as a treatment for syphilis. Its high level of toxicity led to its use as an insecticide. Since arsenic is a semimetallic element, it has properties that make it useful for modern electronics such as transistors.

Antimony c. 1450/1707
Symbol: SB
Atomic number: 51
Atomic weight: 121.75
Melting point: 630.63° C
Boiling point: 1587° C
Atomic radius: 142 pm

Like arsenic, antimony compounds were known in antiquity, but it was not until around 1450 that they were described by Johann Tholden. The name comes from two Greek words, *anti* ("against" or "not") and *monos*, meaning "single" or "solitary," so antimony is "not alone." Antimony's chemical symbol comes from its Latin name, *stibium*. Nicolas Lémery (1645–1715) was the first person to scientifically study antimony and its compounds; he published his work in 1707. Early use of antimony compounds was as a pigment; in 1855, they were used as a component in safety matches. Modern uses include as a coloring for glass and as a trace element in electronics and plastics. Antimony is toxic, so it must be carefully used.

Zinc 1526/1746
Symbol: Zn
Atomic number: 30
Atomic weight: 65.38
Melting point: 419.5° C
Boiling point: 907° C
Atomic radius: 133.2 pm

The name and symbol for zinc comes from Greek *zink,* meaning "prong" or "point." Before zinc was identified as an element, its compounds were used in ancient times, particularly for the manufacturing of brass. As a metal, it was produced in India as early as the thirteenth century. In 1526, the alchemist and physician Paracelsus (1493–1541) identified zinc, but it was not until 1746 that Andreas Sigismund Marggraf (1709–1782) isolated it as an element. The majority of modern zinc use is as a corrosion-resistant coating, such as in galvanized iron, but it is used in thousands of other products, including batteries, paints, CRT screens, cosmetics, and sunscreen. Zinc is also used extensively as an alloy, especially for die casting. It is a necessary trace element in human diet.

Phosphorus 1669/1680   Symbol: P
Atomic number: 15
Atomic weight: 30.974
Melting point: 44.15° C
Boiling point: 280.5° C
Atomic radius: 93 pm

The name "phosphorus" comes from the Greek *phosphoros,* meaning "bringer of light," and was a name for the planet Venus. Phosphorus was discovered by the alchemist Hennig Brand (fl.1670) while he was attempting to created the Philosopher's Stone, but because he kept his method of production secret, it was also independently discovered by Robert Boyle (1627–1691) in 1680. Since phosphorus glows in the dark and easily bursts into flame in warm air, it intrigued researchers. Although phosphorus never occurs in elemental form in nature, its compounds are extremely important, particularly for fertilizer, cleaning agents such as TSP (trisodium phosphate), baking powder, and fine china. Phosphorus is also necessary for life as part of DNA.

Cobalt 1735   Symbol: Co
Atomic number: 27
Atomic weight: 58.9332
Melting point: 1495° C
Boiling point: 2927° C
Atomic radius: 124.3 pm

The name "cobalt" comes from the German *kobalt,* meaning "demon of the mines." This may have come from the fact that this "worthless" mineral often appeared where the miners expected silver. Cobalt compounds have been used since antiquity as a blue pigment, particularly to color glass. Although *kobald* was mentioned by a number of early writers, such as Paracelsus, it was Georg Brandt (1694–1768) who identified it as a unique mineral, around 1735. It was later identified by Lavoisier as an element. In addition to use as pigment, cobalt is used as an alloy for specialized metals such as high-strength magnets and jet-engine steel. The artificially produced cobalt-60 is an important gamma ray producer and is used as a tracer for science experiments and for radiotherapy.

Platinum 1741/1750

Symbol: Pt
Atomic number: 78
Atomic weight: 195.09
Melting point: 1768.4° C
Boiling point: 3825° C
Atomic radius:138 pm

The name and symbol come from Spanish *platina,* meaning "silver." Although this metal was used in pre-Columbian South America, it was identified by a European as a unique mineral in 1557 by Julius Caesar Scaliger (1484–1558). Platinum was described in 1745 by Antonio de Ulloa (1716–1795), but he was prevented from publishing until 1748. Charles Wood independently described the metal in 1741, but it was not identified as a new element until 1750, when it was studied by William Watson (1715–1787). Platinum is very rare and is extensively used as jewelry, but in recent years it has been increasingly used in the electronics industry and as a catalyst, both for scientific research and in commercial applications such as antipollution devices (catalytic converters) and hydrogen fuel cells.

Nickel 1751

Symbol: Ni
Atomic number: 28
Atomic weight: 58.7
Melting point: 1455° C
Boiling point: 2913° C
Atomic radius: 124.6 pm

The name has German and Swedish roots, probably being a contraction of *kopparnickel* or *kupfernickel,* which combined the terms for copper and *nickel,* meaning "demon" or "goblin." Miners found this copper-like mineral, but it produced no copper. Although the metal was known from at least the Middle Ages, it was Axel Fredrik Cronstedt (1722–1765) who isolated and described the metal. He also named it nickel. Nickel exists widely in nature and has been found in meteorites. Almost 30 percent of the world's nickel supply comes from one source, in Sudbury, Ontario, Canada. One of the major uses for nickel is the manufacturing of stainless steel. It is also used in other alloy products, batteries, coinage (hence the term "nickel" for the U.S. five-cent piece), and as a corrosion-resistant coating.

Bismuth 1753

Symbol: Bi
Atomic number: 83
Atomic weight: 208.98
Melting point: 271.4° C
Boiling point: 1564° C
Atomic radius: 155 pm

The name is somewhat obscure, but its symbol came from Latin *bisemutum*, which was a Latinized version of the German *Wismuth*, perhaps from *weisse Masse*, meaning "'white mass." Bismuth, although rarely found in elemental form in nature, may have been confused with tin or lead in ancient times. Although earlier references to the metal may exist, it was Claude Geoffroy Junine who showed that it was a distinct mineral in 1753. Bismuth is used in the creation of magnets, cosmetics, and some medicines and as a trace alloy for specialized iron products.

Magnesium 1755/1808   Symbol: Mg
Atomic number: 12
Atomic weight: 24.305
Melting point: 650° C
Boiling point: 1090° C
Atomic radius: 160 pm

Magnesium was named by Humphry Davy (1778–1829) as the Latinized form of *Magnesia*. Joseph Black (1728–1799) identified *magnesia alba* in 1755, but it was Humphry Davy (1778–1829) who isolated the metal by electrolysis and called it "magnesium" in 1808. Magnesium is the eighth most abundant element on Earth, but it does not occur naturally in elemental form. Pure magnesium burns in warm air with a bright white flame. This property makes it useful for flash bulbs, flares and fireworks, and incendiary bombs. In alloy form, it is used as a lightweight material in airplanes. Traces of magnesium are important for plant and animal life, and it is a component in chlorophylls.

Nitrogen 1772   Symbol: N
Atomic number: 7
Atomic weight: 14.00674
Melting point: −210.0° C
Boiling point: −195.79° C
Atomic radius: 71

Known as "phlogisticated air" and "azote," nitrogen was named by Jean Antoine Chaptal (1756–1832), who proposed the name *nitrogène* in 1790. It combined the Latin *nitrum* (soda) with Greek *genes* (forming). Nitrogen was isolated and described by Daniel Rutherford (1749–1819) in 1772, but it was studied by many of the leading chemists of the time. Nitrogen is very abundant in air, where it forms 78 percent of the atmosphere, and is found as a compound in organic materials. The nitrogen cycle allows life to exist. As a commercial material, nitrogen is used to make ammonia and fertilizer and as a base stock for plastics. Liquid nitrogen is used for cryogenic storage.

Oxygen 1774   Symbol: O
Atomic number: 8
Atomic weight: 15.9994
Melting point: −218.79° C

Boiling point: $-182.95°$ C
Atomic radius: 66 pm

The name and symbol were created in 1789 by Lavoisier, who combined the Greek *oxys* (acid) with *genes* (forming), although his theory of the necessity of oxygen for acid formation was later shown to be incorrect. Although a number of researchers created oxygen, it was first isolated by Carl Wilhelm Scheele (1742–1786) in 1771, but for various reasons his work was not made public until 1777. Joseph Priestley (1733–1804) is generally credited with the first discovery of oxygen (as dephlogisticated air) on August 1, 1774. He published his work soon after and studied its properties for over the next five years. Oxygen is very abundant, forming about 21 percent of the atmosphere and existing in stars, as well as in thousands of compounds. Because of its highly reactive nature, oxygen combines easily with many other elements. It is essential for most forms of life.

Chlorine 1774   Symbol: Cl
Atomic number: 17
Atomic weight: 35.453
Melting point: $-101.5°$ C
Boiling point: $-34.4°$ C
Atomic radius: 99 pm

The name and symbol were created by Humphrey Davy (1778–1829) in 1810. He named the gas after the color, from the Greek *chloros,* meaning "greenish yellow." Chlorine was discovered in 1774 by Carl Wilhelm Scheele (1742–1786). Chlorine does not exist in elemental form in nature, but, because of its reactive nature, it is found in many compounds, particular chloride salts such as NaCl (table salt). Chlorine is one of the most important industrial chemicals. It has many uses, including water purification, bleaching, disinfectant, paints, and textile production. In the form of hydrochloric acid, it is one of the most important industrial chemicals. It was also used as a toxic war gas, particularly in World War I; it is fatal if inhaled in low concentration, but, because of its distinctive odor and color, it was replaced by other chemical weapons that were less detectable.

Manganese 1774   Symbol: Mn
Atomic number: 25
Atomic weight: 54.938
Melting point: 1246° C
Boiling point: 2061° C
Atomic radius: 124 pm

The name and symbol come from the French *manganèse* and Italian *manganese,* from the Latin *Magnesia.* 1774 Carl Wilhelm Scheele characterized the material and called it "manganese." It was isolated as an element by Johan Gottlieb Gahn (1745–1818) in 1774. Manganese occurs widely in nature, but rarely in elemental form. It is used in the manufacturing of magnets and batteries and as a colorant for glass. Traces of manganese give amethysts their color. Manganese dioxide is used in black paint, while permanganate is used for scientific and medical analysis.

Hydrogen 1766

Symbol: H
Atomic number: 1
Atomic weight: 1.0079
Melting point: 259.34° C
Boiling point: −252.87° C
Atomic radius: 78

Named by Lavoisier, the word "hydrogen" combines the Greek *hydro* (water) with *genes* (forming). Although hydrogen had been produced for many years as a by-product of acid action, it was recognized as a distinct substance in 1776 by Henry Cavendish (1731–1810). Hydrogen is the most abundant substance in the universe (perhaps as much as 90 percent of all mass), and all other elements are built up from hydrogen. The commercial use of hydrogen is enormous; it is used in the production of ammonia (fertilizer and chemical base stock) and dirigibles and in the hydrogenation of fats and oils and as a rocket fuel. In the form of hydrochloric acid, it is one of the most important industrial chemicals. There is a great deal of research currently on developing hydrogen as a fuel to replace hydrocarbons. It has the advantage of being nonpolluting (its combustion product is water), but production and storage problems still need to be resolved.

Molybdenum 1778/1781

Symbol: Mo
Atomic number: 42
Atomic weight: 95.94
Melting point: 2623° C
Boiling point: 4639° C
Atomic radius: 136.2 pm

The name and symbol come from the Latin *molybdena* and the Greek *molybdaina*, both meaning "lead ore." Molybdenite ore was confused with both graphite and lead from at least the Middle Ages, but it was recognized as a distinct substance in 1778 by Carl Wilhelm Scheele (1742–1786), who named it. Elemental molybdenum was isolated in 1781 or 1782 by Jacob Hjelm (1746–1813). Molybdenum is usually produced as a by-product of copper or tungsten mining. Molybdenum is used in specialized parts where high-temperature resistance is needed, in specialized electronics, and as a catalyst in petroleum refining. It is also used to produce ultra-high-strength steel. Molybdenum is a necessary trace element in biological processes.

Tellurium 1782/1799

Symbol: Te
Atomic number: 52
Atomic weight: 127.6
Melting point: 449.5° C
Boiling point: 988° C
Atomic radius: 143.2 pm

"Tellurium" comes from the Latin *tellus*, meaning "earth." In 1782, Franz Joseph Müller Freiherr von Reichenstein (1742–1825), a mine inspector in Transylvania, discovered a material that was similar to antimony. He sent it to Martin Heinrich Klaproth (1743–1817), who in 1789 identified it as a new element and named it. Another researcher, Paul Kitaibel (1757–1817), independently identified the new element the same year. Although tellurium occasionally occurs in elemental form, it is generally found as a compound. It is used as an alloy with copper and stainless steel, in some semiconductors, and in the manufacturing of blasting caps.

Tungsten 1779/1781/1783	Symbol: W
	Atomic number: 74
	Atomic weight: 183.5
	Melting point: 3422° C
	Boiling point: 5555° C
	Atomic radius: 137 pm

The name "tungsten" comes from the Swedish *tung sten*, meaning "heavy stone," and was suggested around 1781 by Torbern Bergman (1735–1784). The symbol "W" comes from the convoluted history that led to the discovery of the element. In 1556, Georgius Agricola (1494–1555) noted a material he called *lupi spuma* meaning "wolf's foam," which in German is *wolf rahm*. The mineral ore is today known as wolframite and is one of the main sources of tungsten. In 1779, Peter Woulfe (1727–1803/5) thought he had discovered something new when examining wolframite. This was later confirmed in 1781 by Carl Wilhelm Scheele (1742–1786) and Bergman, but it was not until 1783 that Juan José de Elhuyar y de Zubice and Fausto de Elhuyar y de Zubice (1755–1833) produced the elemental metal. Tungsten occurs only as a compound, and about 75 percent of the known reserves are in China. Tungsten is used extensively as filaments for electric lights and in other electronic equipment. As an alloy, it is used to make strong, heat-resistant steel, particularly for tools and machine parts.

Uranium 1789	Symbol: U
	Atomic number: 92
	Atomic weight: 238.029
	Melting point: 1135° C
	Boiling point: 4131° C
	Atomic radius: 138.5 pm

The name "uranium" and the symbol come from the planet Uranus, discovered in 1781. In turn, Uranus comes from the Greek name *Ouranos*, the son and husband of Gaia. Although uranium oxide had been used as a glass coloring agent from at least 79 C.E., it was Martin Klaproth (1743–1817) who discovered a new element in pitchblende ores in 1789. He named the element after the recently discovered planet. Klaproth did not know that he had found the end of the natural elements, as all heavier elements do not exist in terrestrial form. The elemental metal was isolated in 1841 by Eugène Péligot (1811–1890), and it was found to be radioactive in 1896 by Henri Becquerel (1852–1908). Although uranium is rare, it is widely distributed in the Earth's crust, but only a few places have concentrations that are commercially exploitable. The primary use of uranium is in nuclear power production and in military applications such as nuclear weapons. Because uranium is toxic and radioactive, it can be very dangerous to handle.

Zirconium 1789/1824

Symbol: Zr
Atomic number: 40
Atomic weight: 91.22
Melting point: 1855° C
Boiling point: 4409° C
Atomic radius:160 pm

The name and symbol come from the Persian word *zargûn*, meaning "gold colored." Various forms of zirconium have been known in the form of gemstones since ancient times, but it was Klaproth who noted an unidentified substance in a jargon stone. The elemental metal (in impure form) was isolated in 1824 by Berzelius. Zirconium has been found in stars, meteorites, and lunar rocks. It exists as a compound widely in nature but is usually found in quantities too small to be of commercial interest. Zirconium is used in specialized laboratory equipment, explosives, and lamp filaments and in lotions to treat poison ivy. In the form of zirconium oxide, it is used to make gems and for specialty glass.

Strontium 1787/1791/1808

Symbol: Sr
Atomic number: 38
Atomic weight: 87.62
Melting point: 777° C
Boiling point: 1382° C
Atomic radius: 215.1 pm

The name comes from the town of Strontian in Scotland and was given to the element by Thomas Hope (1766–1844). There are many claims for the original discovery of strontium. William Cruikshank, in 1787, and Adair Crawford, in 1790, both examined strontianite ($SrCO_3$) and recognized that it had unique properties. Thomas Hope noted an unknown "earth" in 1791. Martin Klaproth presented a paper on a number of strontium compounds in 1793 and 1794. Richard Kirwan (1733–1812) examined a number of strontium compounds and presented his findings in 1794. It was Davy who isolated strontium metal, in 1808. Strontium does not occur in pure form in nature but is found in small quantities in many places. Some forms of strontium are radioactive, particularly ^{90}Sr, which has been found in nuclear fallout. It can also be used in SNAP devices (Systems for Nuclear Auxiliary Power) as a power source. The main commercial use of strontium is in the glass of color television picture tubes.

Yttrium 1794/1843

Symbol: Y
Atomic number: 39
Atomic weight: 88.9059
Melting point: 1522° C
Boiling point: 3345° C
Atomic radius: 181 pm

In 1787, the chemist Carl Axel Arrhenius (1757–1824) found a peculiar black rock in the tailings dump of a quarry near the town of Ytterby, Sweden. He named the unknown mineral "ytterbite" after the town. Johan Gadolin (1760–1852) isolated what he thought was a new mineral in 1794, which he called "yttria," but it was Carl Gustav Mosander (1797–1858) who in 1843 isolated yttrium from several other closely related elements in the mineral. Yttrium never occurs in elemental form in nature, but as a compound it is found in trace amounts widely, including lunar rocks and meteorites. The greatest use of yttrium is in the production of red phosphors for color television tubes. It is also used in some high-tech products, such as specialty lasers and rare-earth magnets.

Titanium 1791/1795/1825

Symbol: Ti
Atomic number: 22
Atomic weight: 47.9
Melting point: 1668° C
Boiling point: 3287° C
Atomic radius: 144.8 pm

The name, from the Titans of Greek mythology, was given to the element by Martin Heinrich Klaproth (1743–1817) in 1795. In 1791, William Gregor (1761–1817) discovered a unique mineral in some black sand. Klaproth examined the compounds in the mineral and named the element he believed was one of them. In 1825, J. J. Berzelius isolated an impure form of titanium, while the first pure titanium metal was created in 1910 by Matthew A. Hunter (1878–1961). Titanium, as a compound, is found widely in nature, including in some stars and in lunar rocks. It occurs in most igneous rocks. For many years, titanium was a laboratory curiosity with the strange property of being the only element that burns in nitrogen but having no real use. In 1932, William Kroll (1889–1973) invented a way of mass-producing titanium. Since titanium is as strong as steel but is 45 percent lighter, commercial uses quickly developed, and titanium is found in everything from airplane parts to tennis rackets. The largest use of titanium, however, is as titanium oxide pigment in paint.

Chromium 1797

Symbol: Cr
Atomic number: 24
Atomic weight: 51.996
Melting point: 1907° C
Boiling point: 2671° C
Atomic radius: 124.9 pm

The name and symbol come from the Greek *chroma*, meaning "color." In 1797, Nicolas-Louis Vauquelin (1763–1829) was examining the properties of the mineral crocoite. He found that it turned vibrant colors when mixed with various chemicals. In 1798, he produced chromium metal. Chromium ores are found in a number of places, but it rarely appears in elemental form. The greatest use of chromium is for plating because it produces a hard, reflective surface that resists corrosion. It is also used as an alloy in steel, as a catalyst, and to produce emerald-green glass.

Beryllium 1798/1828   Symbol: Be
Atomic number: 4
Atomic weight: 9.01218
Melting point: 1287° C
Boiling point: 2471° C
Atomic radius: 113.3 pm

The name comes from the Greek *beryllos*, meaning "a sea-green gem." The existence of beryllium was suspected in 1798, when Nicolas-Louis Vauquelin (1763–1829) examined the chemical composition of an emerald, noting some unusual characteristics in the beryl gem and finding that the salts he was producing were sweet tasting. He named the substance "glucina" (sweet). Friedrich Wöhler (1800–1882) prepared the first elemental beryllium in 1828. Martin Heinrich Klaproth (1743–1817) suggested the name "beryllia," since the material seemed like yttria. Beryllium is relatively rare and never occurs in elemental form. Beryllium is the lightest of the metals, and, because of its high melting point and nonmagnetic properties, it has many commercial uses. It is used primarily as an alloy, especially with copper, and as a specialized metal for aerospace parts. It is also used in disc brakes and computer parts, as a moderator for nuclear reactors, and in X-ray lithography of integrated circuits.

Vanadium 1801/1830   Symbol: V
Atomic number: 23
Atomic weight: 50.9414
Melting point: 1910° C
Boiling point: 3407° C
Atomic radius: 132.1 pm

The element was named by Nils Gabriel Sefström (1787–1845) after the Scandinavian goddess of beauty *Vanadis*. In 1801, Andrés Manuel del Rio (1764–1849) identified a new metal he called "erythronium" (from the Greek for red) that seemed related to uranium and chromium, but he was advised that it was actually lead chromate. In 1830, Sefström rediscovered the element and named it, finding out a year later that it was the same material as described by del Rio. Vanadium compounds are found in trace amounts in nature. The majority of vanadium is used as an alloy in steel production.

Niobium (Colombium) 1801/1844

Symbol: Nb
Atomic number: 41
Atomic weight: 92.9064
Melting point: 2477° C
Boiling point: 4744° C
Atomic radius:142.9 pm

The name was given by Heinrich Rose (1795–1864) around 1844 and comes from *Niobe,* the Greek goddess of tears, daughter of *Tantalus.* The history and naming of the element are convoluted. In 1801, Charles Hatchett (1765–1847) found an unknown ore in a rock at the British Museum. It had been sent from North America by the grandson of John Winthrop (1609–1676), the first governor of Connecticut and an alchemist. Hatchett called the new element "columbium" after Christopher Columbus and its American connection. In 1802, Anders Gustaf Ekeberg (1767–1813) claimed to have discovered a new element he called "tantalum." Rose demonstrated that tantalum was not an element but a compound of niobium and what he called "pelopium." It may have been that Hatchett's original columbium was identical to niobium. The elemental metal was first prepared in 1864 by Christian Blomstrand. Although the International Union of Pure and Applied Chemistry officially settled the name issue in 1950, it is still commonly called "columbium." Niobium is used in the aerospace industry and for superconductive magnets and jewelry.

Tantalum 1802/1844/1866

Symbol: Ta
Atomic number: 73
Atomic weight: 180.9479
Melting point: 3017° C
Boiling point: 5458° C
Atomic radius: 143 pm

The name comes from the Greek mythological figure *Tantalos,* one of Zeus's sons. In 1802, Anders Gustaf Ekeberg discovered a new element similar to columbite and named it "tantalum." It was not clear that tantalum and niobium were actually different elements until that was demonstrated by Rowe in 1844 and Marignac in 1866. The pure element was first produced in 1903 by von Bolton. As the history of discovery suggests, tantalum is rare, difficult to isolate, and found with other similar materials. The main use of tantalum is for electrolytic capacitors and vacuum furnace parts and as a trace alloy in high tech equipment. Because it is unaffected by substances in the human body, it is also used in implants such as artificial joints.

Cerium 1803

Symbol: Ce
Atomic number: 58
Atomic weight: 140.12
Melting point: 798° C
Boiling point: 3443° C
Atomic radius: 182.5 pm

The element was named after the newly discovered (1801) asteroid Ceres. Ceres was the Roman goddess of agriculture. The 1803 discovery of cerium is credited to Wilhelm Hisinger (1766–1852), together with Jöns Jakob Berzelius (1779–1848), and to Martin Heinrich Klaproth (1743–1817). Hisinger and Berzelius named the element. Cerium is the most common of the rare-earth elements. It is used in the manufacturing of glass, as a polishing compound, for lighting equipment, and as a catalyst in petroleum refining.

Osmium 1803

Symbol: Os
Atomic number: 76
Atomic weight: 190.2
Melting point: 3033° C
Boiling point: 5012° C
Atomic radius: 135 pm

The name comes from the Greek *osme,* meaning "smell" or "stink," after the smell of the residue from which the element was refined. Osmium was discovered and named in 1803 by Smithson Tennant (1761–1815). Osmium is found in trace amounts with other platinum metals such as rhodium and palladium, and these in turn are found in nickel ores. Osmium metal is primarily used as a hardening alloy for other metals in the platinum group and is thus used in pen nibs, phonograph needles, jewelry, and catalysts.

Palladium 1803

Symbol: Pd
Atomic number: 46
Atomic weight: 106.4
Melting point: 1554.9° C
Boiling point: 2963° C
Atomic radius: 138 pm

Palladium was named in honor of the asteroid Pallas, which was discovered in 1802. Pallas was the Greek goddess of wisdom (also known as Pallas Athena). William Hyde Wollaston (1766–1828) discovered palladium in a sample of ore from South America. Although palladium is used in jewelry (white gold contains palladium), its most important commercial use is as a catalyst in pollution-control devices. Its ability to absorb hydrogen may make it an important element for hydrogen fuel technology.

Iridium 1803

Symbol: Ir
Atomic number: 77
Atomic weight: 192.22
Melting point: 2446° C
Boiling point: 4428° C
Atomic radius: 135.7 pm

The name comes from the Latin *iris,* meaning "rainbow," which in turn comes from *Iris,* the Greek goddess of the rainbow. It was named 1803 by Smithson Tennant (1761–1815), who discovered it while working on ore from

South America. He noticed that it formed compounds with bright colors. A rare element, it is part of the platinum group and is used primarily in platinum products. It is resistance to all acids, so it is also used for laboratory equipment.

Rhodium 1804
Symbol: Rh
Atomic number: 45
Atomic weight: 102.9055
Melting point: 1964° C
Boiling point: 3695° C
Atomic radius: 134.5 pm

Rhodium is named after the Greek *rhodon*, meaning "rose color." William Hyde Wollaston (1766–1828) discovered rhodium in 1804 when he removed platinum and palladium from ore he had received from South America and found a red residue. Rhodium is rare, but it is used commercially in electrical parts and jewelry and as a catalyst.

Potassium (Kalium) 1807
Symbol: K
Atomic number: 19
Atomic weight: 39.098
Melting point: 63.5° C
Boiling point: 759° C
Atomic radius: 227 pm

The name comes from the English *potash*, while the symbol comes from the Latin *kalium* and the Arabic *qali*, from which we also derive the word "alkali." Although potash (wood ashes) had been used for a variety of products since antiquity, it was Humphry Davy who isolated potassium in 1807. Because the metal came from potash, he named it potassium. Because of its similarity to an existing word in German, Martin Heinrich Klaproth and Ludwig Wilhelm Gilbert (1769–1824) in 1813 suggested the alternative *kalium*, and it is from this that the symbol is derived. It was the first element produced by electrolysis. Potassium minerals are abundant and make up almost 2.5 percent of the mass of the Earth's crust. Potash is primarily used as a fertilizer, while potassium salts are used in thousands of products, from food to photography.

Sodium (Natrium) 1807
Symbol: Na
Atomic number: 11
Atomic weight: 22.98977
Melting point: 97.8° C
Boiling point: 883° C
Atomic radius: 144.4

The English name sodium was created by Humphry Davy (1778–1829) because he isolated sodium metal from soda by electrolysis in 1807. Soda was a version of the medieval Latin *sodanum*, a name for a headache remedy, which in turn probably came from the Arabic *Sudâ* (soda). Because of its similarity to an existing word in German, Martin Heinrich Klaproth and Ludwig Wilhelm Gilbert (1769–1824), in 1813, suggested "natrium," from the Latin *nitron,* meaning soda. It is from this term that the symbol is derived. Sodium in various forms is widely distributed in nature and is found in the stars, notable because of the D lines found in solar spectra. Sodium metal is used to produce esters and in specialty alloys. Sodium chloride (NaCl—table salt) is the most common form of sodium and is used in thousands of products, particularly food production and preservation. Sodium is necessary for life.

Calcium 1808
Symbol: Ca
Atomic number: 20
Atomic weight: 40.08
Melting point: 842° C
Boiling point: 1484° C
Atomic radius: 197.3 pm

Calcium was named by Humphry Davy after the material from which it was produced, *calx,* the Latin for limestone. Davy isolated calcium metal in 1808 by electrolysis. Although calcium never occurs in elemental form, calcium compounds are widely found in nature, particularly in limestone and gypsum. Calcium metal is used in the production and purification of other metals, while calcium compounds have thousands of uses, including chemical production, plaster, and Portland cement.

Barium 1808
Symbol: Ba
Atomic number: 56
Atomic weight: 137.34
Melting point: 727° C
Boiling point: 1897° C
Atomic radius: 217.3 pm

The name and symbol come from Greek *barys,* meaning "heavy." The existence of a new element was suspected in 1772 when Carl Wilhelm Scheele (1742–1786) noticed some small crystals in a sample of pyrolusite. He named the mineral "baryta." Barium was isolated by electrolysis in 1808 by Humphry Davy, who named it following the existing root. Barium is relatively rare and occurs only in combination with other elements. Elemental barium has limited uses, but barium compounds are found in paint and rat poison; it is also used as a coloring agent in fireworks and in medical research and glassmaking. Some isotopes of barium are radioactive, and barium was significant in the discovery of nuclear fission.

Boron (Bore) 1808
Symbol: B
Atomic number: 5
Atomic weight: 10.81
Melting point: 2075° C
Boiling point: 4000° C
Atomic radius: 83 pm

Both versions of the name comes from Arabic *Bauraq* and Persian *burah*, referring to what is now called borax. In 1808, Louis Joseph Gay-Lussac (1778–1850) and Louis-Jacques Thénard (1777–1857), in France, and Humphry Davy, in England, isolated the element from borax. Gay-Lussac and Thénard suggested "bore," while Davy suggested the term "boracium." The English name was shorted to boron because it was related to carbon. Although boron does not exist in elemental form in nature, its compounds are relatively common. Boron is used in fireworks and rockets, while boron compounds are important for the manufacturing of fiberglass, bleach, textiles, and glass.

Iodine 1811/1813
Symbol: I
Atomic number: 53
Atomic weight: 126.9045
Melting point: 113.7° C
Boiling point: 184.4° C
Atomic radius: 133.3 pm

The name and symbol come from the Greek *iodes*, meaning "violet." Iodine was discovered in 1811 by Bernard Courtois (1777–1838), but it was identified as an element by Charles-Bernard Désormes (1777–1862) and his partner Nicolas Clément (1779–1841) in 1813. Humphry Davy suggested the name "iodine," while Gay-Lussac suggested "iode." Iodine never occurs in elemental form in nature. Its compounds are used in organic chemistry and medicine, primarily as a disinfectant. Traces of iodine are necessary for good health, and the lack of iodine produces goiter (a swelling of the thyroid gland); table salt is iodized for that reason.

Lithium 1817
Symbol: Li
Atomic number: 3
Atomic weight: 6.941
Melting point: 180.5° C
Boiling point: 1342° C
Atomic radius: 152 pm

The name and symbol come from the Greek *lithos*, meaning "stone." In 1817, Johan August Arfvedson (1792–1841) discovered lithium in a sample of petalite. He named it "lithion" because it came from a mineral source. This name was later transformed into lithium to match other names. Lithium is the lightest of all the metals, but because it is corrosive and highly reactive, it does not occur in elemental form. Trace amounts of lithium can be found in most igneous rock. It is used commercially in alloys, nuclear applications, and batteries. Lithium carbide is used as a treatment for various mental illnesses.

Cadmium 1817
Symbol: Cd
Atomic number: 48
Atomic weight: 112.4
Melting point: 321.07° C
Boiling point: 767° C
Atomic radius: 148.9 pm

The name comes from Latin *cadmia,* referring to the mineral calamine. The discovery of cadmium is one of the most complex stories from the periodic table. At least nine different people claim to have discovered the new element in 1817 or 1818. In 1817, a sample of "suspect" zinc oxide was examined by J.C.H. Roloff, an inspector of pharmacies. The producer was Carl Samuel Hermann (1765–1846), who began to work on the material. At the same time, the team of Kluge and Staberoh, a pair of medical assessors who were examining part of Roloff's sample, declared they had found a new metal they named "Klaprothium," after the German chemist Martin Heinrich Klaproth, who had died the year before. Friedrich Stromeyer (1776–1835), inspector general of pharmacies, also examined the material and found the new metal. He named it "kadmium." Other claims came from W. Meissner, Dr. von Vest (perhaps Lorenz von Vest der Jüngere), and J.F.W. Brandes. Cadmium does not exist in elemental form in nature and is relatively rare. It is primarily used in electroplating and Ni-Cd batteries. Cadmium phosphors are used for blue and green color in television tubes.

Selenium 1817

Symbol: Se
Atomic number: 34
Atomic weight: 78.96
Melting point: 220.5° C
Boiling point: 685° C
Atomic radius: 117 pm

Named after *Selene,* the Greek name for the moon, selenium was isolated by Jakob Berzelius (1779–1848) in 1817. Because its properties were similar to those of tellurium (named after the Earth), Berzelius named it after the moon. Selenium is rare and never occurs in elemental form. It is used in photocopiers, electronic components, and glassmaking and as an additive in specialty stainless steel.

Silicon (Silicium) 1824

Symbol: Si
Atomic number: 14
Atomic weight: 28.086
Melting point: 1414° C
Boiling point: 3265° C
Atomic radius: 117 pm

The name comes from the Latin *silex,* meaning "flint." The name "silicium" was the original suggested name, having the same root. "Silicum" is used outside the English-speaking world. Although it is likely that Humphry Davy came close to isolating silicon in 1800, it was Jakob Berzelius (1779–1848) who first prepared amorphous silicon in 1824 and identified it as an element. Silicon is the second most abundant element in the Earth's crust and has been found in stars and meteorites. Although it is never found in elemental form, its oxide and silicates are everywhere. Silicon as a compound has a huge range of uses, from sand for glassmaking to clay for bricks and pottery, and it is the basis for electronics such as transistors and other solid-state devices. Flint was one of the most important natural materials for human history in early toolmaking, and it was later used with steel as a way to make fire.

Aluminum (Aluminium) 1825/1827

Symbol: Al
Atomic number: 13
Atomic weight: 26.98154
Melting point: 660.32° C
Boiling point: 2519° C
Atomic radius: 143.1 pm

The name is derived from the Latin *alumen,* which was the name for *alum,* a material that had been used as a medicine and an ingredient for the dye industry since ancient times. It was suspected as early as 1761 by Louis-Bernard Guyton de Morveau (1737–1816) that there was an unknown metallic base in alum. In 1787, Antoine Lavoisier suggested that alumine was the oxide of the unknown metal. In 1807, Sir Humphry Davy (1778–1829) proposed the name "aluminum," and shortly after, the name "aluminium" was adopted by IUPAC to conform with the -ium ending of most elements. Both terms are now in general use, although North Americans tend to use "aluminum." In 1825, Hans Christian Ørsted (1777–1851) was the first person to prepare metallic aluminum, but he did not follow up on the work. In 1827, Friedrich Wöhler (1800–1882) developed a method to isolated the metal and became known as the discoverer of aluminum. Although aluminum is widely found in nature, making up about 8 percent of the Earth's crust, it is found only in combination with other materials and must be smelted to produce useable metal. It is used anywhere strong, light-weight materials are desirable, from pots to car engines.

Bromine 1826

Symbol: Br
Atomic number: 35
Atomic weight: 79.904
Melting point: −7.2° C
Boiling point: 58.8° C
Atomic radius: 115 pm

The name comes from the Greek *bromos,* meaning "stench," and was suggested by the French Academy of Science because of bromine's irritating odor. In 1826, Antoine-Jérôme Balard (1802–1876) found evidence of an unknown substance in seaweed, while at the same time Carl Löwig (1803–1890), working with chlorine, found the same thing in the residue of an experiment. Balard published first, getting priority. Bromine is the only nonmetallic liquid element. It is not found in elemental form in nature and is primarily extracted from brine. It is used in fumigants, photography, and dyes and as an alternative to chlorine in water purification systems, such as those used in swimming pools.

Thorium 1828   Symbol: Th
Atomic number: 90
Atomic weight: 232.0381
Melting point: 1750° C
Boiling point: 4788° C
Atomic radius: 179.8 pm

Named after *Thor,* the Scandinavian god of war, thorium was discovered by Jöns Jakob Berzelius (1779–1848) in 1828 in a mineral sample from Norway. Although thorium does not occur in elemental form, thorium ores are thought to be about as abundant as lead in the Earth's crust. Thorium has limited uses as a commercial product, primarily in specialty electronics and Welsbach mantles (portable gas lights). Because thorium is radioactive and more plentiful than uranium, it may be used as a power source in the future. The internal heating of the Earth may be a result of the action of thorium.

Lanthanum 1839   Symbol: La
Atomic number: 57
Atomic weight: 138.9055
Melting point: 918° C
Boiling point: 1897° C
Atomic radius: 187.7 pm

The name was suggested in 1839 by Jöns Jakob Berzelius (1779–1848) and is from the Greek word *lanthanein,* meaning "to be concealed or lie hidden." Lanthanum was discovered in 1839 by Carl Gustav Mosander (1797–1858) when he was working on cerium compounds. Although lanthanum is not found in elemental form, lanthanum oxides are used in lighting equipment, particularly for movies. They are also used in specialty optical glass. Certain lanthanum alloys can reversibly absorb hydrogen, making them potentially useful for hydrogen energy systems.

Terbium 1843   Symbol: Tb
Atomic number: 65
Atomic weight: 158.9254
Melting point: 1356° C
Boiling point: 3230° C
Atomic radius: 178.2 pm

The name comes from a truncation of the name Ytterby, a town in Sweden where the original ore was discovered. It was one of a number of rare-earth elements linked to yttrium and was first identified by Carl Gustav Mosander (1797–1858) in 1843. It was not prepared in elemental form until 1905. It is rare and hard to refine. Terbium has few uses and is used primarily as a compound in some solid-state electronics.

Erbium 1843   Symbol: Er
Atomic number: 68
Atomic weight: 167.26

Melting point: 1529° C
Boiling point: 2868° C
Atomic radius: 175.7

The name comes from a truncation of the name Ytterby, a town in Sweden where the original ore was discovered. It was one of a number of rare-earth elements linked to yttrium and was first identified by Carl Gustav Mosander (1797–1858) in 1843. It was prepared in impure elemental form in 1905 and in pure form in 1934. One of the rare earths associated with yttria, it is a relatively rare compound and has few uses. Erbium oxide has been used to give glass a pink color.

Ruthenium 1844   Symbol: Ru
Atomic number: 44
Atomic weight: 101.07
Melting point: 2334° C
Boiling point: 4150° C
Atomic radius: 134 pm

The name comes from *Ruthenia,* Latin for "Russia." In 1840, Karl Karlovich Klaus (1796–1864) started an investigation to settle a dispute between Gottfried Wilhelm Osann (1797–1866) and Jöns Jakob Berzelius (1779–1848) over the existence of new elements in residue from platinum ore found in the Ural Mountains. In 1844, Klaus identified and named the new element. Small quantities of ruthenium are found in platinum and nickel ores. It is used primarily as a catalyst and a hardening alloy for platinum products.

Cesium (Caesium) 1860   Symbol: Cs
Atomic number: 55
Atomic weight: 132.9054
Melting point: 28.5° C
Boiling point: 671° C
Atomic radius:265.4 pm

The name comes from the Latin *caesius,* meaning "sky blue." Cesium was discovered by Robert Wilhelm Bunsen (1811–1899) and Gustav Robert Kirchhoff (1824–1887) in 1860. They used a spectroscope on a drop of mineral water and saw previously unnoted blue lines in the spectra. Cesium is rare, but it is used in photoelectric cells and as a hydrogenation catalyst. It is also used in some atomic clocks.

Rubidium 1861   Symbol: Rb
Atomic number: 37
Atomic weight: 85.4678
Melting point: 39.3° C
Boiling point: 688° C
Atomic radius: 247.5 pm

The name comes from the Latin *rubidus*, meaning "deep red." Rubidium was discovered by Gustav Robert Kirchhoff (1824–1887) and Robert Wilhelm Bunsen (1811–1899) in 1861, using their spectroscope. They named it after the red lines found in the spectra of the new element. It is rare, and it is radioactive. It is used in photoelectric cells and specialty glass. An exotic compound of rubidium, silver, and iodine may be useful in thin film batteries.

Thallium 1861

Symbol: Tl
Atomic number: 81
Atomic weight: 204.37
Melting point: 303.5° C
Boiling point: 1473° C
Atomic radius: 170.4 pm

The name comes from *thallos*, Greek for "green twig." It was discovered by William Crookes using spectroscopic analysis, who named it after the color of the spectra line. The metal was isolated by Crookes and independently by Claude-Auguste Lamy (1820–1878) in 1862. The elemental metal does not occur naturally and is extracted as a by-product from refining pyrites, lead, or zinc. In its pure form it is shiny but oxidizes quickly. It resembles lead. The element is toxic and should be handled carefully. It primary commercial use is in rodent poison and ant killer, as well as in photo cells.

Indium 1863

Symbol: In
Atomic number: 49
Atomic weight: 114.82
Melting point: 156.6° C
Boiling point: 2072° C
Atomic radius: 162.6 pm

The name is derived from the Latin word *indicum*, referring to an indigo pigment. It was named after the bright indigo spectral line discovered in 1863 by Ferdinand Reich (1799–1882) and Hieronymus Theodor Richter (1824–1898). Metallic indium was first produced in 1924. It is a silvery-white metal. About 4 million ounces worldwide are produced annually. The elemental form does not occur naturally. Indium is used as a trace element in electronics and specialty alloys and as a substitute for silver in the manufacturing of mirrors.

Gallium 1875

Symbol: Ga
Atomic number: 31
Atomic weight: 69.72
Melting point: 29.76° C
Boiling point: 2204° C
Atomic radius: 122.1 pm

The element is named using the Latin name *Gallia*, for France. It was predicted and described in 1863 by Mendeleev as eka-aluminum and discovered in 1875 by Paul Émile (François) Lecoq de Boisbaudran (1838–1912) using spectroscopic analysis. He isolated elemental gallium the same year. The element occurs as a trace element in a number of minerals, including coal, but does not exist in pure form. Because gallium has such a low melting point and high boiling point, it is used in high-temperature thermometers. It is also used for mirrors and as a doping agent for electronics.

Holmium 1879   Symbol: Ho
Atomic number: 67
Atomic weight: 164.9304
Melting point: 1474° C
Boiling point: 2700° C
Atomic radius: 176.6 pm

The name is from the Latin *Holmia* for the city of Stockholm, the region where the original minerals were found. The discovery of holmium was part of the complex yttria rare-earths research. A new element was suggested by Jacques-Louis Soret (1827–1890) in 1878 and was named by Per Theodor Cleve (1840–1905) in 1880; its oxide was isolated in 1886 by Lecoq de Boisbaudran. The pure metal was isolated in 1911 by O. Holmberg. It is a rare element and does not occur in pure form in nature. The silvery metal has unusual magnetic properties but no current commercial uses.

Thulium 1879   Symbol: Tm
Atomic number: 69
Atomic weight: 168.9342
Melting point: 1545° C
Boiling point: 1950° C
Atomic radius: 174.6 pm

The element is named after Thule, the ancient name for Scandinavia. It is part of the yttria rare earths and is about as abundant as silver or cadmium. It was discovered in 1879 by Per Theodor Cleve. The silvery metallic element does not occur in elemental form naturally. It is very expensive and has few commercial uses, although synthetic radioactive isotopes have been used as an X-ray source.

Scandium 1879   Symbol: Sc
Atomic number: 21
Atomic weight: 44.9559
Melting point: 1541° C
Boiling point: 2836° C
Atomic radius: 160.6 pm

The name is derived from the Latin *Scandia*, the name of Scandinavia. Mendeleev predicted the existence of the element as eka-boron. It was discovered by Lars Fredrick Nilson (1840–1899) in 1878 when he was investigating trace elements in euxenite and gadolinite. Metallic scandium was first prepared in 1937. Spectroscopic analysis has shown that scandium is present in the sun and other stars. Blue beryl may be colored because of scandium. The silvery metallic element is used in high-intensity lighting.

Ytterbium 1879   Symbol: Yb
                              Atomic number: 70
                            Atomic weight: 173.04
                             Melting point: 819° C
                            Boiling point: 1196° C
                            Atomic radius: 194 pm

The element is named after Ytterby, the village in Sweden were the original minerals the element came from were found. In 1878, Jean Charles Galissard de Marignac (1817–1894) and Carl Auer von Welsbach (1858–1929) both identified the trace element, although Marignac's name was accepted. A purer form of the element was prepared in 1937, but it was not until 1953 that pure metallic ytterbium was produced and could be analyzed. There are currently few uses for the silvery metallic element, although it may have applications in the manufacturing of specialty steel, and a radioactive isotope may be used as an X-ray source.

Samarium 1880   Symbol: Sm
                              Atomic number: 62
                            Atomic weight: 150.4
                            Melting point: 1074° C
                            Boiling point: 1794° C
                            Atomic radius: 180.2 pm

Samarium is named after the mineral samarskite, from which the first traces were found. The history of its discovery is very convoluted, and it is part of the complex yttria group. The first suggestion of a new element came in 1846, but it was not until 1878 that J. Lawrence Smith (1818–1883) announced a new element he called "mosandrum." His discovery was disputed, and, in 1879, François Lecoq de Boisbaudran named it "samaria," which was later changed to "samarium." The whitish-silver element does not occur in elemental form in nature. It has been used in lighting systems and also as part of rare-earths magnets, in lasers, in infrared absorbing glass, and in nuclear reactors as a neutron buffer.

Gadolinium 1880   Symbol: Gd
                              Atomic number: 64
                            Atomic weight: 157.25
                            Melting point: 13131° C
                            Boiling point: 3273° C
                            Atomic radius: 180.2 pm

The name comes from gadolinite, the mineral from which the trace element was first found. It was identified in 1880 by both Jean Charles Galissard de Marignac (1817–1984) and François Lecoq de Boisbaudran (1838–1912) and is part of the complex yttria group. The whitish-silver metal has been used in compound form in specialty alloys, magnets, and phosphors in color television.

Praseodymium 1885

Symbol: Pr
Atomic number: 59
Atomic weight: 140.9077
Melting point: 931° C
Boiling point: 3520° C
Atomic radius: 182.8 pm

The name combines the Greek word *prasios,* meaning "green," and *didymos,* meaning "twin." The name partly refers to the green color of the oxide. Part of the complex yttria group of elements, it was discovered by Carl Auer von Welsbach (1858–1929) in 1885 when he separated two elements from didymium (so named because it closely resembled lanthanum). These were neodymium (new twin) and praseodymium. It was prepared in relatively pure form in 1931. It is a scarce material and does not exist in pure form in nature. There are few commercial uses, but it has been used in studio lighting and as an coloring additive to glass and enamel.

Neodymium 1885

Symbol: Nd
Atomic number: 60
Atomic weight: 144.24
Melting point: 1021° C
Boiling point: 3074° C
Atomic radius: 182.1 pm

The name comes from the combination of the Greek *neos,* meaning "new," and *didymos,* meaning "twin." Part of the complex yttria group of elements, it was discovered by Carl Auer von Welsbach (1858–1929) in 1885 when he separated two elements from didymium (so named because it closely resembled lanthanum). These were neodymium and praseodymium (green twin). It was isolated in relatively pure form in 1925, but it was long used as an alloy (about 18%) known as misch metal (containing cerium 50% and lanthanum 25%). Misch metal is used in lighters. The metallic element does not occur naturally. Other than as a spark source for lighters, it is used as a colorant in glass and enamels, producing a violet to wine-red tint.

Germanium 1886

Symbol: Ge
Atomic number: 32
Atomic weight: 72.59
Melting point: 938.25° C
Boiling point: 2833° C
Atomic radius: 122.5 pm

The name comes from the Latin *Germania,* for Germany. In "The Periodic Law of the Chemical Elements," in 1869, Mendeleev predicted the existence of the element as eka-silicon. It was discovered by Clemens Alexander Winkler (1838–1904) in 1886 and named by him. Germanium metal is whitish-gray and is refined from a number of ores, including zinc and coal. It is one of the most commercial important of the rare-earth elements as a doping element for a wide range of semiconductors. It is also used in specialty glass.

Fluorine 1886   Symbol: F
Atomic number: 9
Atomic weight: 18.998
Melting point: −219° C
Boiling point: −188° C
Atomic radius: 70.9 pm

The name comes from the Latin *fluere,* meaning "to flow." The ore from which the element was extracted was described as early as 1529 by Georigius Agricola as fluorspar. It was named by Humphry Davy, but it was not isolated until 1886 by Ferdinand-Frédéric-Henri Moissan (1852–1907). Fluorine is the most electronegative and reactive of all elements. At standard temperature and pressure, it is a pale yellow gas, and, because it is so reactive and poisonous, it is difficult to handle. Its first large-scale use was by the Manhattan Project when uranium hexafluoride gas was created to extract uranium-235 for the atomic bomb. Hydrofluoric acid is used to etch glass, and chlorofluorocarbons (CFCs) were once used extensively in air conditioning, refrigeration, and aerosol propellant until they were was banned as being dangerous to the ozone layer in the 1990s. As fluoride, it is used to control cavities, often as an additive to municipal water supplies.

Argon 1894   Symbol: Ar
Atomic number: 18
Atomic weight: 39.948
Melting point: −189.35° C
Boiling point: −185.85° C
Atomic radius: 174 pm

The name comes from the Greek *argos,* meaning "inactive." The existence of argon was suspected by Cavendish in 1785, but it was not identified until 1894 by Sir William Ramsay (1852–1916) and Lord Rayleigh (1842–1919) and announced in 1895. It was the first of the "noble gases," referred to as such because they do not combine with other "common" elements under normal circumstances. It required the addition of a column to the periodic table. Argon is extracted from atmospheric air, where it makes up about 1 percent of the gases present. Argon is a colorless gas, and its greatest commercial use is in light bulbs and lighting systems. It is also used as an inert gas shield for arc welding and high-temperature cutting.

Helium 1895   Symbol: He
Atomic number: 2
Atomic weight: 4.0026
Melting point: −272.2° C
Boiling point: −268.93° C
Atomic radius: 128 pm

The name comes from the Greek *helios,* meaning "the sun." It was first detected by its spectra by Pierre-Jules-César Janssen (1824–1907) in the corona of the sun during a solar eclipse in 1868. It was named by Norman Lockyer and Edward Frankland (1825–1899) in 1869. William Ramsay (1852–1916) demonstrated that helium could be found on Earth in 1895, when he detected it in gas given off by the heating of the mineral cleveite. Helium is the second most abundant element in the universe, in large part because it is the first stage product of hydrogen fusion and therefore part of the energy cycle of stars. The study of helium has been extensive, especially in relationship to cryogenic research. At normal pressure, helium will liquefy, but it does not solidify even at absolute zero. As one of the noble gases, it does not normally combine with other elements. As the second lightest gas, it is used for lighter than air craft and dirigibles, where it is safer than hydrogen. It is used commercially to inflate party balloons, in deep-sea diving, as an inert gas for arc welding, and as a coolant for laboratory equipment and other devices.

Krypton 1898　　　　Symbol: Kr
Atomic number: 36
Atomic weight: 83.8
Melting point: −157.38° C
Boiling point: −153.22° C
Atomic radius: 189 pm

Krypton is named after the Greek *kryptos,* meaning "hidden." It was named by William Ramsay (1852–1916) and Morris W. Travers (1872–1961), who were looking for other trace gases in the atmosphere. They were looking for a gas lighter than argon but discovered the spectral traces of krypton (brilliant green and orange lines), a heavier gas. Trace amounts of krypton are found in atmospheric air. It is used in chemical analysis and in some specialty photographic flash lamps.

Neon 1898　　　　Symbol: Ne
Atomic number: 10
Atomic weight: 20.179
Melting point: −258.59° C
Boiling point: −246.08° C
Atomic radius: .848 pm

Named for the Greek *neos,* meaning "new," neon was discovered by William Ramsay (1852–1916) and Morris W. Travers (1872–1961) in 1898 as part of their search for a noble gas between helium and argon. Neon is a trace component of atmospheric air. Its main use is in electric signs, where it glows red. It is also used in lasers, lightning arrestors, and televisions. Liquid neon is used for cryogenic research and as a refrigerant.

Xenon 1898　　　　Symbol: Xe
Atomic number: 54
Atomic weight: 131.3
Melting point: −111.79° C
Boiling point: −108.12° C
Atomic radius: 218 pm

The name comes from the Greek *xenon,* meaning "stranger." Xenon was discovered by William Ramsay (1852–1916) and Morris W. Travers (1872–1961) in 1898 as part of their search for a noble gas between helium and argon. It is present as a trace element in atmospheric air. It is the heaviest of the noble gases. It is used commercially in specialty lamps and lasers, as well as in sophisticated laboratory equipment such as bubble chambers and as a radioactive isotope used as a tracer.

Polonium 1898

Symbol: Po
Atomic number: 84
Atomic weight: 209
Melting point: 254° C
Boiling point: 962° C
Atomic radius: 167 pm

The name comes from the Latin *Polonia,* meaning Poland, the home country of Marie Curie, one of its discoverers. It was discovered by Marie Curie (1867–1934) and Pierre Curie (1859–1906) in 1898 when they were studying uranium and other radioactive materials found in pitchblende. Polonium is very rare, and, although some exists naturally, most polonium is manufactured in nuclear reactors. Polonium is very dangerous even in minute quantities because of its level of radioactivity. It is a very good source of alpha radiation and, if combined with beryllium, produces neutrons. It is thus used as a thermoelectric source for specialized applications such as satellites.

Radium 1898

Symbol: Ra
Atomic number: 88
Atomic weight: 226.0254
Melting point: 700° C
Boiling point: 1140° C
Atomic radius: 223 pm

The name comes from the Latin *radius,* meaning "ray." It was discovered by Marie and Pierre Curie in 1898 when they were studying uranium and other radioactive materials found in pitchblende. There is about 1 g of radium in 7 tons of pitchblende, but it is 3×10^5 times more radioactive than uranium. It was isolated as a metallic element in 1911 by Marie Curie and André-Louis Debierne (1874–1949). Radium exists in small quantities associated with uranium ores. Radium is phosphorescent, so it has been used to make luminous paint, especially for watch dials, but, because it is highly radioactive, most uses are related to nuclear medicine or the energy industry. Radon gas is produced from radium and is a harmful by-product.

Actinium 1899

Symbol: Ac
Atomic number: 89
Atomic weight: 227

> Melting point: 1051° C
> Boiling point: 3159° C
> Atomic radius: 187.8 pm

The name comes from the Greek *aktis*, meaning "beam" or "ray." It was discovered by André-Louis Debierne (1874–1949) in 1899 and independently by Fritz Giesel (1852–1927) in 1902. It exists in very small quantities in association with uranium ores. Actinium has few uses outside the laboratory, but its discovery was important for the development of chemistry and physics, as it was one of the materials used to study radioactive decay, since it breaks down into thorium, radium, radon, bismuth, polonium, and isotopes of lead.

> **Radon 1899**   Symbol: Rn
> Atomic number: 86
> Atomic weight: 222
> Melting point: −71° C
> Boiling point: −61.7° C
> Atomic radius: 120 pm

The name is derived from the term "radium emanation" and was suggested in 1923 as a means of clarifying and simplifying the naming of radioactive materials. In 1899, Robert B. Owens (1870–1936?) noted problems measuring the radiation of thorium. With Ernest Rutherford, they speculated that there was a radioactive gas being produced. The first controlled production of radon was by Marie and Pierre Curie, with Friedrich Ernst Dorn (1848–1916) confirming the findings in 1900. Radon at standard temperature and pressure is a colorless gas and is generally inert. It can be found in trace amount in the atmosphere. Commercially, radon is used in nuclear medicine. Because it is a by-product of the radioactive decay of naturally occurring elements such as radium and thorium, it can accumulate in mines and pose a health hazard.

> **Europium 1901**   Symbol: 63
> Atomic number: Eu
> Atomic weight: 151.96
> Melting point: 822° C
> Boiling point: 1529° C
> Atomic radius: 204.2 pm

Named after the continent of Europe, europium is part of the complex yttria family of rare earths. It was isolated in 1901 by Eugène-Anatole Demarçay (1852–1904), who also named it. Europium is one of the rarest and most expensive rare-earth elements. In its pure form, it is silvery white and resembles lead. It is used as part of the red phosphor in color televisions and as a doping element for lasers.

> **Lutetium 1907**   Symbol: Lu
> Atomic number: 71
> Atomic weight: 174.97
> Melting point: 1663° C

Boiling point: 3401° C
Atomic radius: 173.4 pm

The name comes from the Latin *Lutetia Parisorum*, meaning "the city of Paris." In 1907, Georges Urbain (1872–1938) isolated what he called "lutecia" from ytterbia, one of the rare-earth compounds that had previously been considered an element. In 1949, IUPAC modified the spelling. Because Auer von Welsbach was also working on these materials, the element is sometimes known as "cassiopeium." Lutetium is part of the complex yttria family of rare earths. Lutetium emits beta radiation and has been used as in a limited fashion as a catalyst for hydrogenation and other chemical processes.

Protactinium 1917   Symbol: Pa
Atomic number: 91
Atomic weight: 231.0359
Melting point: 1572° C
Boiling point: ~4000° C
Atomic radius: 160.6

The name combines the Greek *protos*, meaning "first," with the element actinium (89). It was named by Otto Hahn (1879–1968) and Lise Meitner (1878–1968) in 1918. Element 91 was predicted by Mendeleev in 1871 eka-tantalum. It was discovered by Kasimir Fajans (1887–1975) and Otto H. Göhring, who called it "brevium,'" meaning "brief," because it decayed quickly. Protactinium is one of the rarest and most expensive naturally occurring elements; 125 g of protactinium was produced from 60 tons of nuclear waste material by the Great Britain Atomic Energy Authority in 1961, and this remains the largest stock of the material. It has scientific uses but no commercial uses.

Hafnium 1922   Symbol: Hf
Atomic number: 72
Atomic weight: 178.49
Melting point: 2233° C
Boiling point: 4603° C
Atomic radius: 156.4 pm

The name comes from *Hafnia*, the Latin name for Copenhagen. The element was named by Dirk Coster (1889–1950) and György Karl von Hevesy (1885–1966), who used X-ray spectroscopy to demonstrate its existence. The story of the discovery of hafnium is complex, since an unknown element associated with zircon had been suspected since at least 1845 and claims, counterclaims, and rejections of claims followed. Names including norium (1845), ostranium (c.1845), jargonium (1869), norwegium (1879), nigrium (1911), euxenium (1911), and celtium (1911) were all suggested but not adopted because the evidence was not strong enough. When Niels Bohr announced Coster and von Hevesy's work as part of his Nobel Prize acceptance speech, it confirmed their name and priority of discovery. Elemental hafnium is brilliant silver and has characteristics similar to those of zircon. It is used for control rods in nuclear reactors and in some lighting systems.

Rhenium 1908/1925

Symbol: Re
Atomic number: 75
Atomic weight: 186.207
Melting point: 3186° C
Boiling point: 5596° C
Atomic radius: 137.0 pm

The name is from the Rhineland in Germany. In 1908, the Japanese chemist Masataka Ogawa (1865–1933) discovered this element and called it "nipponium," but he incorrectly assigned it to periodical space 43. In 1925, the research team of Walter Noddack (1893–1960), Ida Eva Tacke (1896–1978, later Ida Noddack), and Otto Berg (1873–?) extracted 1 g of rhenium from molybdenite ore and correctly characterized its properties. Rhenium is relatively rare. Its main uses are in laboratory equipment such as spectrographs, in filaments for photoflash lamps, and in some specialty electrical equipment.

Francium 1939

Symbol: Fr
Atomic number: 87
Atomic weight: 223
Melting point: 27° C
Boiling point: 677° C
Atomic radius: 270 pm

The element's name comes from the country France, and it was named in 1946 by Marguerite Perey (1909–1975), who discovered the element in 1939. Element 87 was predicted in 1871 by Mendeleev. He gave it the name ekacaesium. It was known first as actinium-K as a radioactive product of the decay of actinium, and it is the most unstable of the first 101 elements. Although francium occurs naturally, its short half-life means that there are only a few grams of the element at any time in the crust of the Earth. Because the longest lasting isotope of francium lasts only 22 minutes, it has no commercial uses.

Technetium 1937

Symbol: Tc
Atomic number: 43
Atomic weight: 97
Melting point: 2157° C
Boiling point: 4265° C
Atomic radius: 135.8 pm

The name comes from the Greek *technetos*, meaning "artificial." There were claims to have discovered element 43 as early as 1877, based on the predicted existence of eka-manganese. In 1925, Walter Noddack, Ida Tacke, and Otto Berg claimed to have found the element, calling it "masurium." Definitive proof was made by Emilio G. Segrè (1905–1989) in 1937, who was studying a small piece of radioactive molybdenum created at the Berkeley cyclotron.

Technium was the first synthetic element, although a naturally occurring isotope was discovered in minute quantities in 1962. The metal is silvery grey. It has not been used as a commercial product but may have uses in specialty steel and superconducting devices.

Neptunium 1940

Symbol: Np
Atomic number: 93
Atomic weight: 237.0482
Melting point: 644° C
Boiling point: 3900° C
Atomic radius: 131 pm

The name comes from *Neptunus,* the Latin name for the god of the sea, but it was named after the planet Neptune, which had recently been discovered. The element was first prepared in 1940 by Edwin M. McMillan and Philip Abelson at the Berkeley Laboratory of the University of California. They irradiated uranium with neutrons to create the new element. Neptunium does not exist in nature and is primarily of scientific interest. It is used in neutron detection equipment.

Astatine (astatium) 1940

Symbol: At
Atomic number: 85
Atomic weight: 210
Melting point: 302° C
Boiling point: ~340° C
Atomic radius: 145 pm

The name is from the Greek *astatos,* meaning "unstable." Mendeleev predicted the existence of element 85, calling it eka-iodine. Although there were a number of prior claims, the element was first produced in 1940 by Dale R. Corson (1914–), Kenneth R. Mackenzie (1912–2002), and Emilio Segrè (1905–1989). They synthesized the isotope ^{211}At by irradiating bismuth with alpha particles. Although there is some naturally occurring astatine, like francium, it is constantly produced by radioactive decay of other elements and decays itself. Astatine has no commercial uses and is primarily of scientific interest in studies of radioactivity.

Plutonium 1940

Symbol: Pu
Atomic number: 94
Atomic weight: 244
Melting point: 640° C
Boiling point: 3228° C
Atomic radius: 131 pm

Named after the planet Pluto, which is named after the Greek god Plouton, the ruler of the underworld, plutonium was first produced in 1940 by the team of Glenn T. Seaborg, Edwin M. McMillan, Joseph W. Kennedy, and Arthur C. Wahl. They used the 60-inch cyclotron at Berkeley, California, to bombard uranium with deuterons. Although a very small amount of plutonium exists in nature most plutonium is produced in breeder reactors. Because plutonium is highly fissionable, it was produced and used as one of the energy sources for the atomic bomb dropped on Nagasaki. Plutonium is used for power production and for nuclear weapons.

Americium 1944	Symbol: Am
	Atomic number: 95
	Atomic weight: 243
	Melting point: 1176° C
	Boiling point: 2011° C
	Atomic radius: 184 pm

The element is named for the continent of North America. The element was created by Glenn T. Seaborg (1912–1999), Ralph A. James, Leon O. (Tom) Morgan, and Albert Ghiorso. The team synthesized americium by bombarding plutonium with neutrons. Americium is an intense gamma source and therefore is very dangerous to handle. It is used as a source for gamma radiography, in various gauges, and as an ionization source for smoke detectors.

Curium 1944	Symbol: Cm
	Atomic number: 96
	Atomic weight: 247
	Melting point: 1345° C
	Boiling point: 3100° C
	Atomic radius: 170 pm

Named after Marie and Pierre Curie, curium was first prepared in 1944 by the team of Glenn T. Seaborg, Ralph A. James, and Albert Ghiorso at the Metallurgical Laboratory in Chicago as part of the Manhattan Project. They used helium ion bombardment of plutonium to produce the element. Very small amounts of curium may exist in the Earth's crust as a result of decay and neutron capture in uranium, but the only usable amounts are made synthetically. Although curium has no commercial uses at present, with its high thermal activity, it might be used as a power source in the future.

Promethium 1945	Symbol: Pm
	Atomic number: 61
	Atomic weight: 145
	Melting point: 1042° C
	Boiling point: ~3000° C
	Atomic radius: 181 pm

The name comes from the Greek mythological figure Prometheus, who stole fire from the gods and gave it to man. The existence of element 61 was predicted in 1902 by Bohuslav Brauner, who extended the periodic table downward after lanthanum. He suggested that there would be an element between neodymium and samarium. Attempts to find the element proved to be inconclusive until it was isolated in 1945 by the team of Charles D. Coryell, Jacob (Jack) A. Marinsky, Lawrence E. Glendenin, and Harold G. Richter. They identified promethium in the by-products of uranium fission. Promethium has not been found on Earth, but its spectral pattern has been identified in at least one star. It has been used as a source of beta radiation for thickness gauges, and it has been proposed for nuclear batteries.

Berkelium 1949

Symbol: Bk
Atomic number: 97
Atomic weight: 247
Melting point: 1050° C
Boiling point: ?
Atomic radius: ?

Named after Berkeley, California, berkelium was created in 1949 by the team of Stanley G. Thompson, Albert Ghiorso, and Glenn T. Seaborg at the Berkeley Laboratory of the University of California. They bombarded americium with helium ions to create the new element. Elemental berkelium has not been synthesized, and only microgram quantities of berkelium chloride have been made, but enough to characterize the element. Primarily of scientific interest, it has no commercial applications.

Californium 1949

Symbol: Cf
Atomic number: 98
Atomic weight: 251
Melting point: 860° C
Boiling point: ?
Atomic radius:~186

Named after the state of California, californium was first created in 1950 by the team of Stanley G. Thompson, Kenneth Street, Jr., Albert Ghiorso, and Glenn T. Seaborg at the Berkeley Laboratory of the University of California. They bombarded curium with helium ions to produce the element. No californium exists terrestrially. It is a strong neutron source and has been used in specialized detection gauges and mineral assay equipment.

Einsteinium 1954

Symbol: Es
Atomic number: 99
Atomic weight: 252
Melting point: 860° C
Boiling point: ?
Atomic radius: ~186 pm

Named after Albert Einstein, Eensteinium was detected as a by-product of a test of a thermonuclear weapon. The test, codenamed "Mike," was conducted on Eniwetok Atoll in the Pacific Ocean in November 1952. The element was first synthesized in 1952 (although its existence was kept secret until 1955 for security reasons) by the team of Albert Ghiorso, Stanley G. Thompson, Gary H. Higgins, Glenn T. Seaborg, Martin H. Studier, P. R. Fields, Sherman M.

Fried, H. Diamond, J. F. Mech, G. L. Pyle, John R. Huizenga, A. Hirsch, W. M. Manning, C. I. Browne, H. Louise Smith, and R. W. Spence, who irradiated uranium with neutrons. In 1961, enough einsteinium was created to evaluate its properties. Einsteinium is primarily of scientific interest and has no commercial applications.

Fermium 1954	Symbol: Fm
	Atomic number: 100
	Atomic weight: 257
	Melting point: 1527° C
	Boiling point: ?
	Atomic radius: ?

Named after Enrico Fermi, fermium was detected as a by-product of a test of a thermonuclear weapon. The test, codenamed "Mike," was conducted on Eniwetok Atoll in the Pacific Ocean in November 1952. Minute quantities (about 200 atoms) were extracted from coral at the test site. It was first synthesized in 1953 (although its existence was kept secret until 1955 for security reasons) by the team of Albert Ghiorso, Stanley G. Thompson, Gary H. Higgins, Glenn T. Seaborg, Martin H. Studier, P. R. Fields, Sherman M. Fried, H. Diamond, J. F. Mech, G. L. Pyle, John R. Huizenga, A. Hirsch, W. M. Manning, C. I. Browne, H. Louise Smith, and R. W. Spence, who bombarded uranium with oxygen ions. The element is primarily of scientific interest, and there are no commercial applications.

Mendelevium 1955	Symbol: Md
	Atomic number: 101
	Atomic weight: 258
	Melting point: 827° C
	Boiling point: ?
	Atomic radius: ?

Named after Dimitri Mendeleev, mendelevium was first synthesized in 1955 by the team of Albert Ghiorso, Bernard G. Harvey, Gregory R. Choppin, Stanley G. Thompson, and Glenn T. Seaborg. They used the 60-inch cyclotron at the Berkeley Laboratory of the University of California to bombard einsteinium with helium ions. Only enough atoms of mendelevium have been made to confirm its identity.

Nobelium 1958	Symbol: No
	Atomic number: 102
	Atomic weight: 259
	Melting point: 827° C
	Boiling point: ?
	Atomic radius: ?

Nobelium is named after Alfred Nobel. Two American laboratories and one Russian one claimed the creation of element 102, but the 1958 synthesis by the team of Albert Ghiorso, Torbjørn Sikkeland, J. R. Walton, and Glenn

T. Seaborg at the Lawrence Radiations Laboratory, University of California, was later judged to be the first verified production. The researchers created nobelium by bombarding curium with carbon ions. Only minute quantities of nobelium have been produced, and it is of scientific interest only. The longest lasting isotope has a half-life of three minutes.

Lawrencium 1961

Symbol: Lr
Atomic number: 103
Atomic weight: 262
Melting point: 1627° C
Boiling point: ?
Atomic radius: ?

Named after Ernest O. Lawrence, lawrencium was first synthesized in 1961 by the team of Albert Ghiorso, Torbjørn Sikkeland, Almon E. Larsh, and R. M. Latimer at the Berkeley Laboratory of the University of California. They bombarded californium with boron ions. The longest lasting isotope of lawrencium has a half-life of 35 seconds, just long enough to test its oxidation rate and to confirm its status as an element.

Rutherfordium (unnilquadium) 1964

Symbol: Rf
Atomic number: 104
Atomic weight: 261
Melting point: ?
Boiling point: ?
Atomic radius: ?

Named after Ernest Rutherford, element 104 was the subject of a naming dispute, since the element was first synthesized in 1964 by the Russian team of Georgy Nikolaevich Flerov, Yuri Tsolakovich Oganessian, Y. V. Lobanov, V. I. Kuznetsov, V. A. Druin, V. P. Perelygin, K. A. Gavrilov, S. P. Tretiakova, and V. M. Plotko at the Joint Institute for Nuclear Research (JINR) at Dubna. They created it by bombarding plutonium with neon ions and named the element "kurchatovium." In 1969, the American team of Albert Ghiorso, M. Nurmia, J. Harris, K. Eskola, and P. Eskola at the Berkeley Laboratory of the University of California also created element 104 by irradiation of californium with carbon ions. They proposed the name "rutherfordium," and it was this name that was accepted by IUPAC in 1997. Because the half-life of rutherfordium is so short, in the five-second range, its characteristics are inferred primarily by its production and decay products.

Dubnium (unnilpentium) 1968

Symbol: Db
Atomic number: 105
Atomic weight: 262
Melting point: ?
Boiling point: ?
Atomic radius: ?

Named for Dubna, the town were the laboratory that discovered the element is located, the element was created in 1968 by the Russian team of Georgy Nikolaevich Flerov, V. A. Druin, A. G. Demin, Y. V. Lobanov, Nikolai Konstantinovich Skobelev, G. N. Akap'ev, B. V. Fefilov, I. V. Kolesov, K. A. Gavrilov, Yu. P. Kharitonov, and L.P. Chelnokov at the Joint Institute for Nuclear Research (JINR) at Dubna. The researchers synthesized the new element by bombarding americium with neon ions. They proposed the named "nielsbohrium." In 1970, the American team of Albert Ghiorso, M. Nurmia, K. Eskola, J. Harris, and P. Eskola at the Berkeley Laboratory of the University of California also created element 105 by bombarding californium with nitrogen ions. They suggested the name "hahnium" after Otto Hahn. In 1994, IUPAC suggested "joliotrium," but in 1997 the element was officially named dubnium. Given dubnium's maximum half-life of 40 seconds, only its basic characteristics have been verified.

Seaborgium (unnilhexium) 1974

Symbol: Sg
Atomic number: 106
Atomic weight: 263
Melting point: ?
Boiling point: ?
Atomic radius: ?

Named after Glenn Seaborg, element 106 was first prepared in 1974 by the team of Albert Ghiorso, J. M. Nitschke, J. R. Alonso, C. T. Alonso, M. Nurmia, E. Kenneth Hulet, R. W. Lougheed, and Glenn T. Seaborg at the Berkeley Laboratory of the University of California. It was also created at the Joint Institute for Nuclear Research (JINR) at Dubna. A priority and naming dispute arose over the creation of element 106. The American team bombarded californium with oxygen ions, while the Russian group irradiated lead with chromium ions. Joint credit for discovery was given to both laboratories, while the Americans felt they had priority. In 1994, the American group proposed the name "'seaborgium," while IUPAC suggested "rutherfordium." In 1997, IUPAC recognized "seaborgium" as the accepted name. The half-life of seaborgium is under one second. Its characteristics are inferred from analysis of its fission products.

Bohrium (unnilseptium) 1976/1981

Symbol: Bh
Atomic number: 107
Atomic weight: 262
Melting point: ?
Boiling point: ?
Atomic radius: ?

The element is named after Neils Bohr. The first announced detection of element 107 was at the Joint Institute for Nuclear Research (JINR) in Dubna. Because the experimenters "saw" the element for only 2/1000th of a second, it was difficult to determine anything about it. The first confirmation of the existence of element 107 came in 1981 and is credited to the team of Gottfried Münzenberg, Sigurd Hofmann, Fritz Peter Heßberger, Willibrord Reisdorf, Karl-Heinz Schmidt, J.R.H. Schneider, W.F.W. Schneider, Peter Armbruster, Christoph-Clemens Sahm, and B. Thuma at the Gesellschaft für Schwerionenforschung in Darmstadt, Germany. The name "bohrium" was recognized by IUPAC in 1997.

Meitnerium (unnilenium) 1982

Symbol: Mt
Atomic number: 109
Atomic weight: 265
Melting point: ?
Boiling point: ?
Atomic radius: ?

Named after Lise Meitner, element 109 was first prepared in 1982 by the team of Gottfried Münzenberg, Peter Armbruster, Fritz Peter Heßberger, Sigurd Hofmann, Klaus Poppensieker, Willibrord Reisdorf, K. Schneider, Karl-Heinz Schmidt, Christoph-Clemens Sahm, and Detlef Vermeulen at the Gesellschaft für Schwerionenforschung in Darmstadt, Germany. They bombarded bismuth with iron nuclei. After a week of bombardment, they produced one atom of element 109. The name "meitnerium" was accepted by IUPAC in 1997.

Hassium (unniloctium) 1984

Symbol: Hs
Atomic number: 108
Atomic weight: 265
Melting point: ?
Boiling point: ?
Atomic radius: ?

Named after the Latin *Hassias* for the German state of Hessen, hassium was synthesized in 1984 by the team of Gottfried Münzenberg, Peter Armbruster, H. Folger, Fritz Peter Heßberger, Sigurd Hofmann, J. Keller, Klaus Poppensieker, Willibrord Reisdorf, Karl-Heinz Schmidt, H.-J. Schött, Matti Leino, and R. Hingmann at the Gesellschaft für Schwerionenforschung in Darmstadt, Germany.

Darmstadtium (ununnilium) 1994

Symbol: Ds
Atomic number: 110
Atomic weight: ?
Melting point: ?
Boiling point: ?
Atomic radius: ?

Named after Darmstadt, Germany, element 110 was sought by a number of laboratories, but the first atom was created in 1994 by the team of Sigurd Hofmann, Victor Ninov, Fritz Peter Heßberger, Peter Armbruster, H. Folger, Gottfried Münzenberg, and H. J. Schött at the Gesellschaft für

Schwerionenforschung in Darmstadt, Germany. Also working on the element were Andre Georgievich Popeko, Alexander Vladimirovich Yeremin, and A. N. Andreyev, at the Flerov Laboratory of Nuclear Reactions in Dubna, Russia; S. Saro and Rudolf Janik at Katedra jadrovej fyziky, Univerzita Komenského, in the city of Bratislava, Slovakia; and Matti Leino, of the Fysiikan laitos, in Jyväskylän Yliopisto, Finland. Darmstadtium was created by fusing a nickel and lead atom together. This fusion lasted less than 1/1000th of a second. It is expected that more stable isotopes of the element will be discovered.

Unununium (roentgenium) 1994

Symbol: (Rg)
Atomic number: 111
Atomic weight: ?
Melting point: ?
Boiling point: ?
Atomic radius: ?

This name was proposed to honor Wilhelm Conrad Röntgen. The element was first synthesized in 1994 by the team of Sigurd Hofmann, Victor Ninov, Fritz Peter Heßberger, Peter Armbruster, H. Folger, Gottfried Münzenberg, and H. J. Schött, at the Gesellschaft für Schwerionenforschung in Darmstadt, Germany; Andre Georgievich Popeko, Alexander Vladimirovich Yeremin, and A. N. Andreyev, at the Flerov Laboratory of Nuclear Reactions in Dubna, Russia; S. Saro and Rudolf Janik, of the Katedra jadrovej fyziky, Univerzita Komenského, in the city of Bratislava, Slovakia; and Matti Leino, of the Fysiikan laitos, in Jyväskylän Yliopisto, Finland. It was created by the irradiation of bismuth with nickel nuclei. The name has not yet by officially recognized by IUPAC, but it is already appearing in print.

Ununbium 1994

Symbol: uub
Atomic number: 112
Atomic weight: 277
Melting point: ?
Boiling point: ?
Atomic radius: ?

No proposed name has been accepted. As of 2003, the unambiguous identification of element 112 had not been established. It was first synthesized in 1994 by the team of Sigurd Hofmann, Victor Ninov, Fritz Peter Heßberger, Peter Armbruster, H. Folger, Gottfried Münzenberg, and H. J. Schött at the Gesellschaft für Schwerionenforschung in Darmstadt, Germany; Andre Georgievich Popeko, Alexander Vladimirovich Yeremin, and A. N. Andreyev, at the Flerov Laboratory of Nuclear Reactions in Dubna, Russia; S. Saro and Rudolf Janik, of the Katedra jadrovej fyziky, Univerzita Komenského, in the city of Bratislava, Slovakia; and Matti Leino, of Fysiikan laitos, in Jyväskylän Yliopisto, Finland. The element was created by fusing calcium with uranium. Its half life was 280 microseconds.

Ununquadium 1999   Symbol: uuq
Atomic number: 114
Atomic weight: ?
Melting point: ?
Boiling point: ?
Atomic radius: ?

No proposed name has been accepted. As of 2003, the unambiguous identification of element 114 had not been established. The element was synthesized by Yuri Tsolakovich Oganessian, Alexander Vladimirovich Yeremin, Andre Georgievich Popeko, Sergey L. Bogomolov, German Vladimirovich Buklanov, M. L. Chelnokov, Victor Ivanovich Chepigin, Boris N. Gikal, Vladimir A. Gorshkov, Georgy G. Gulbekian, Michael Grigorievich Itkis, Anatoly Petrovich Kabachenko, Anton Yurievich Lavrentiev, Oleg Nikolaevich Malyshev, Jozef Rohach, and Roman Nikolaevich Sagaidak at the Flerov Laboratory of Nuclear Reactions in Dubna, Russia. It was created by the fusing of calcium with plutonium. The observed half-life of 30 seconds seems to confirm that there may be "islands of stability" around element 114.

THE ELEMENTS BY ALPHABETICAL ORDER

Name	Earliest Date	Name	Earliest Date
actinium	1899	chromium	1797
aluminum	1827	cobalt	1735
americium	1944	copper	antiquity
antimony	c.1450	curium	1944
argon	1894	darmstadtium	1994
arsenic	c.1250	dubnium	1968
astatine	1940	einsteinium	1954
barium	1808	erbium	1843
berkelium	1949	europium	1901
beryllium	1798	fermium	1954
bismuth	1753	fluorine	1886
bohrium	1976	francium	1939
boron	1808	gadolinium	1880
bromine	1826	gallium	1875
cadmium	1817	germanium	1886
calcium	1808	gold	antiquity
californium	1949	hafnium	1922
carbon	antiquity	hassium	1984
cerium	1803	helium	1895
cesium	1860	holmium	1879
chlorine	1774	hydrogen	1766

Name	Earliest Date	Name	Earliest Date
indium	1863	(roentgenium)	1994
iodine	1811	rhenium	1908
iridium	1803	rhodium	1804
iron	antiquity	rubidium	1861
krypton	1898	ruthenium	1844
lanthanum	1839	rutherfordium	1964
lawrencium	1961	samarium	1880
lead	antiquity	scandium	1879
lithium	1817	seaborgium	1974
lutetium	1907	selenium	1817
magnesium	1755	silicon	1824
manganese	1774	silver	antiquity
mendelevium	1955	sodium	1807
meitnerium	1982	strontium	1787
mercury	antiquity	sulfur	antiquity
molybdenum	1778	tantalum	1802
neodymium	1885	technetium	1939
neon	1898	tellurium	1782
neptunium	1940	terbium	1843
nickel	1751	thallium	1861
nitrogen	1772	thorium	1828
niobium (columbium)	1801	thulium	1879
		tin	antiquity
nobelium	1958	titanium	1791
osmium	1803	tungsten	1779
oxygen	1774	(ununbium)	(1994)
palladium	1803	(ununquadium)	(1999)
phosphorus	1669		
platinum	1735	uranium	1789
plutonium	1940	vanadium	1801
polonium	1898	xenon	1898
potassium	1807	ytterbium	1879
praseodymium	1885	yttrium	1794
		zinc	1526
promethium	1945	zirconium	1789
protactinium	1917		
radium	1898		
radon	1899		

THE ELEMENTS BY DATE OF DISCOVERY

Date	Name	Date	Name
Antiquity	carbon	1778	molybdenum
	copper	1779	tungsten
	gold	1782	tellurium
	iron	1789	uranium
	lead	1789	zirconium
	mercury	1791	titanium
	silver	1794	yttrium
	sulfur	1797	chromium
	tin	1798	beryllium
c.1250	arsenic	1801	niobium
c.1450	antimony	1830	vanadium
1526	zinc	1802	tantalum
1669	phosphorus	1803	cerium
1735	cobalt	1803	iridium
1735	platinum	1803	osmium
1751	nickel	1803	palladium
1766	hydrogen	1804	rhodium
1774	chlorine	1807	potassium
1774	manganese	1807	sodium
1774	oxygen	1808	barium
1775	magnesium	1808	boron

Date	Name	Date	Name
1808	calcium	1900	radon
1808	strontium	1901	europium
1811	iodine	1907	lutetium
1817	cadmium	1908	rhenium
1817	lithium	1917	protactinium
1818	selenium	1922	hafnium
1824	silicon	1939	francium
1826	bromine	1939	technetium
1827	aluminum	1940	astatine
1828	thorium	1940	neptunium
1839	lanthanum	1940	plutonium
1843	erbium	1944	americium
1843	terbium	1944	curium
1844	ruthenium	1945	promethium
1860	cesium	1949	berkelium
1861	rubidium	1949	californium
1861	thallium	1954	einsteinium
1863	indium	1954	fermium
1875	gallium	1955	mendevium
1879	holmium	1958	nobelium
1879	ytterbium	1961	lawrencium
1879	scandium	1964	rutherfordium
1879	thulium	1968	dubnium
1880	gadolinium	1974	seaborgium
1880	samarium	1976	bohrium
1885	neodymium	1982	meitnerium
1885	praseodymium	1984	hassium
1886	fluorine	1994	darmstadtium
1886	germanium	1994	(roentgenium)
1894	argon	(1994)	(ununbium)
1895	helium	(1999)	(ununquadium)
1898	krypton		
1898	neon		
1898	polonium		
1898	radium		
1898	xenon		
1899	actinium		

GLOSSARY

Alchemy: The study and manipulation of matter based on a belief that matter can be transformed from one form to another by physical, spiritual or magical processes, or a combination of such processes. Although alchemical ideas extend back to at least 5000 B.C.E and existed in all civilizations, there was a great expansion of alchemical activities between 800 and 1600 C.E. after which time alchemical ideas were gradual replaced with empirical concepts of matter. Alchemy was often directed toward the making of gold from base metals or the creation of an elixir of life that gave immortality.

Alloy: A mixture of two or more substances, usual made to gain specific characteristics such as hardness or resistance to corrosion. Bronze is an alloy of copper and tin.

Alpha particles: The same as helium nuclei.

Atom: The word comes originally from the Greek word *atomos*, meaning "uncut" or "indivisible." The ancient Greek philosophers known as the Epicureans believed that matter was composed on small, hard particles they called *atomos*. The idea and the word was revived in the 1600s to represent the smallest fundamental particle of matter. Today it means the smallest part of an element that can exist independently and interact chemically.

Atomic number: Represents the number of protons in the nucleus of an atom and is thus also equal to the number of electrons in orbit around a neutral atom of an element.

Atomic weight: A number representing the mass of an atom. This is a relative scale that compares the average mass of atoms of any element to 1/12 of the mass of a carbon-12 atom. The atomic weight number is the only one in science that does not have a symbol such as °C for temperature or kg for weight. Also known as atomic mass or relative atomic mass.

Base metal: Any metal that is not one of the precious metals. The term "base" means "inferior or impure." Historically, gold and silver were the only precious metals people recognized, but today the elements silver, gold, platinum, palladium, iridium, gallium, rhodium, osmium, and ruthenium are all considered precious metals by the U.S. National Institute for Standards and Technology.

Beta particles: Electrons not tied to a nucleus.

Caloric: Introduced by Antoine Lavoisier and Pierre Laplace as the substance of heat. It allowed an empirical measurement of heat but was discarded in favor of the kinetic theory of heat. We retain the term "calorie" as a measurement of energy, particularly food energy.

Calx: A metal oxide formed by the exposure of metal to oxygen.

Corpuscle: The smallest particle. Although the terms "corpuscle" and "atom" seem interchangeable, they were not, since there were many thinkers who visualized matter as composed of small particles but who did not believe that those particles were elemental or indivisible.

Corpuscularian: One who believed in matter composed of corpuscles.

Electron: One of the three basic subatomic components of all elements. It is a negatively charged particle with virtually no mass that orbits the nucleus of an atom.

Element: The most fundamental form of matter at a chemical level. The great range of ideas about elements and what substances should be considered elements has driven chemical research for much of the history of civilization. Early thinkers generally assumed that there was a small number of elements (from 1 to 5), but today we recognize about 110 naturally occurring and artificial elements found on the **Periodic Table of Elements**.

Elixir: An alchemical term referring to a drug or method of extending life or gaining immortality. Many alchemists, particularly in the Indian and Chinese traditions, were more interested in the Elixir of Life than in the creation of gold from lead.

Energeticist: One who held that all physical phenomenon, including the apparent existence of matter, could be explained in terms of fields of energy.

Ether: The term "ether" has been used to indicate a number of different substances. It was used to indicate the matter of the celestial realm, making up the stars and the planets and the substance through which they traveled. It was seen as a unique element that did not exist on Earth. The idea of ether was revived in various ways to fill in the space between hard particles in systems that did not allow the existence of empty space or a vacuum. The term was also used in the late 1800s to explain the passage of light and radio waves through space, but experiments failed to reveal the existences of the ether, and it was eventually dropped in favor of Einstein's model of space-time. Not to be confused with the modern term "ether," used to indicate compounds that link an oxygen atom with two alkyl groups, such as diethyl ether (CH_3-CH_2-O-CH_2-CH_3).

Fission: The splitting of a large atomic nucleus into two or more smaller parts. This is usually the result of a collision between a fast-moving neutron and a nucleus. When the nucleus breaks apart, a small amount of matter is converted to energy. This is the basis for nuclear power plants and for the first nuclear weapons.

Fusion: The melding of two or more nuclei to form a larger nucleus. In the great temperature and pressure of the Sun, hydrogen nuclei undergo fusion to produce helium atoms. In the process, a small amount of mass is converted to energy, providing the energy of the Sun. The same process produces the destructive power of the hydrogen bomb.

Heat: The energy in a substance represented by the motion of the particles in the substance. For most of history, fire (and later heat) was considered a separate substance.

Imponderable fluid: A substance that cannot be weighed or otherwise measured but whose existence can be inferred from chemical activity. **Phlogiston** and **caloric** were imponderable fluids. The concept was discarded as unscientific in the 1800s.

Isotope: One of two or more versions of an atom of the same element that have the same number of protons but different numbers of neutrons. This makes them chemically similar but physically different. For example, hydrogen has three forms: hydrogen (1 proton, no neutron); deutronium (1 proton, 1 neutron); and tritium (1 proton, 2 neutrons). Although all elements have isotopes, the isotopes of radioactive elements are very important for everything from nuclear energy to carbon dating.

Molecule: A collection of two or more atoms that are bonded together to form a chemical substance. Almost all terrestrial matter is in the form of molecules.

Natural philosophy: A subdiscipline of philosophy that studies the natural world and seeks to provide a complete system of description and purpose for nature. It is often associated with Greek and medieval philosophy. It is a precursor to modern science, but natural philosophers often rejected the use of experiments as a method of proof, one of the key ideas in modern science.

Neutron: One of the three basic subatomic components of all elements except hydrogen. It is a massive particle that has no electrical charge in the nucleus of an atom. Most hydrogen atoms consist of only an electron and a proton.

Noble gases: Sometimes also known as "inert gases" because of their general low lever of chemical activity. The noble gases are helium, neon, argon, krypton, xenon or radon, and ununoctium. The name comes from the same root as noble metals, which did not interact chemically with many other substances, were associated with rarity and wealth and were often owned only by the nobility.

Noble metal: The same as **precious metal.** These metals (general gold and silver) did not interact chemically with many other substances and were associated with rarity and wealth and thus were often owned only by the nobility.

Oxidization: The chemical combination of oxygen with a metal. The process of oxidation was one of the great mysteries of chemistry.

Particle: An indefinite term for a small (often the smallest possible) portion of matter.

Periodic Table of Elements: A listing of the elements arranged by atomic number. The modern table comes most directly from the work of Dimitri Mendeleev but has been modified since the discovery of the subatomic components of the atom.

Philosopher's Stone: An alchemical term named after Aristotle. Some alchemists believed that the Philosopher's Stone was an actual object that could be made and had the power to cause the **transmutation** of base matter such as lead into precious matter like gold. Other alchemists used the term allegorically, referring to the process of transmutation rather than to a physical object that would cause such a change.

Phlogiston: An **imponderable fluid** that was the essence of fire. This special type of material had fluid-like characteristics, so it would flow from regions of high concentration to regions of low concentration. Phlogiston was used as a way of explaining combustion, respiration, **calx** formation (rusting or **oxidization**), and the heat of living things.

Prime element: A type of matter that was undifferentiated. In other words, it was not one of the terrestrial elements or compounds. In alchemy, the prime element could be removed from matter through purification, and then it could become anything, such as gold, by being exposed to the pattern of existing matter.

Proton: One of the three basic subatomic components of all elements. It is a massive, positively charged particle in the nucleus of an atom.

Quintessence: Literally, meaning "fifth element." It was used to describe the matter of the celestial realm and was also used as a synonym for **prime matter.**

Radioactivity: The spontaneous breakdown of the nucleus of an atom through the emission of particles (**alpha** and **beta** particles) and gamma radiation. Although all atomic nuclei can theoretically disintegrate, under normal conditions only a few of the naturally occurring elements such as uranium and radium undergo any significant rate of decay.

Spectroscope: A laboratory instrument invented by Robert Bunsen and Gustav Kirchhoff in 1859. Using a colorless flame, the spectroscope first heats a sample of matter until it releases light. The light is passed through a prism to produce a spectrum. Each element has a unique set of spectral colors, so the device can be used to identify elements and elements in compounds. The spectroscope was also used to identify a number of previously unknown elements, such as helium. The spectroscope is still used today, and astronomical spectroscopes are used to identify stars and even to tell their age.

Temperature: A measurement of the heat energy in a substance. There are three main temperature scales used in the world: Celsius, Fahrenheit, and Absolute (also called the Kelvin scale).

Transmutation: The idea that matter can be made to change from one form to another. Alchemists generally wanted to transmute base metals such as lead or iron into gold or silver. After the discovery of radioactivity, it was found that one element could be converted to another, such as uranium to lead. The fusion process in stars could be considered a kind of cosmic transmutation.

Ultramicroscope: A special form of microscope invented by Richard Zsigmondy and H. Siedentopf in 1903. It uses the Tyndall effect (light scattering from small particles) to make observations of particles at the absolute limit of optical resolution at about 5 millimicrons. Used in experiments that demonstrated the existence of physical molecules and atoms.

Wave: Periodic motion in or through a medium. Historically, there has been a great deal of work done on waves, particularly light waves. In modern physics, anything that moves (including atoms or electrons within atoms) can be described in the mathematics of waves.

ANNOTATED BIBLIOGRAPHY

Interest in the development of chemistry has been on the rise in the past 10 years after a period of scanty publication in the 1970s and 1980s. As a result, this bibliography includes some outstanding material that was published more than 30 years ago but that is still considered standard knowledge for people interested in the field of history of science and chemistry, as well as some very recent material that has started to bring new perspectives to the history of chemistry. The older material can usually be found in libraries and sometimes in used bookstores. The Internet has proven a wonderful tool for tracking down older material. Also, in this electronic age, there are many Internet sources that have useful and interesting information about the history of chemistry, the story of the elements, and up-to-the-minute news about these topics.

The history of the elements fits within the larger subject of the history of science. One of the most entertaining new books is Bill Bryson's *A Short History of Nearly Everything*. Full of great stories, it covers a number of the most important episodes in the development of science. A more comprehensive overview of the history of science can be found in Andrew Ede and Lesley Cormack's *A History of Science in Society* (2004). Other well-known surveys are J. D. Bernal's *Science in History* (1969) and Stephen F. Mason's *A History of the Sciences* (1962). These works can be found in larger libraries but are now out of print. It is interesting to note that both authors were chemists by profession before turning to history. Heilbron's *The Oxford Companion to the History of Science* (2003) is a collection of material on particular topics, including the Chemical Revolution of the eighteenth century.

The concept of the elements and the search for them has been at the heart of chemistry since the beginning, so histories of chemistry contain a great deal of material about the elements. Excellent overviews of chemistry can be found in William H. Brock's *The Norton History of Chemistry* (1992), Aaron J. Ihde's

The Development of Modern Chemistry (1964), and Bernard Jaffe's *Crucibles: The Story of Chemistry from Ancient Alchemy to Nuclear Fission* (1976). There are paperback versions of all of these books. A newer work that includes the advances in chemistry of the past 30 years is Trevor Levere's *Transforming Matter: A History of Chemistry from Alchemy to the Buckyball* (2001).

J. R. Partington was one of the founding fathers of the history of chemistry. His *A Short History of Chemistry* (1989) is an excellent introduction, while his encyclopedic four-volume *A History of Chemistry* (1961–1970) is a reference work that covers almost every topic in chemical history.

Specific interest in the periodic table and the elements has produced a number of recent books. One of the first serious examinations of the history of the elements was done by May Elvira Weeks. *Discovery of the Elements* (1968), an updated version with material from Henry M. Leicester, can still be found in libraries. Richard Morris's *The Last Sorcerers: The Path from Alchemy to the Periodic Table* (2003) is an excellent book, while Paul Strathern's *Mendeleyev's Dream* (2000) is a nontechnical look at the hunt for order among the elements that reads almost like a novel. More technical material on matter theory can be found in Antio Clericuzio, *Elements, Principles, and Corpuscles: A Study of Atomism and Chemistry in the Seventeenth Century* (2000); David M. Knight, *Atoms and Elements. A Study of Theories of Matter in England in the Nineteenth Century* (1967); and Mary Jo Nye, *From Chemical Philosophy to Theoretical Chemistry: Dynamics of Matter and Dynamics of Disciplines 1800–1950* (1993).

Books that contain some historical material but look mostly at the chemical properties and uses (if any) of the elements include Albert Stwertka, *A Guide to the Elements* (2002); John Emsley, *Nature's Building Blocks: An A-Z Guide to the Elements* (2003); and Philip Ball, *The Elements: A Very Short Introduction* (2004).

For anyone who wants to know what the scientists actual wrote about their work, a collection of important original documents can be found in Henry M. Leicester and Herbert S. Klickstein's *A Source Book in Chemistry, 1400–1900* (1968). In addition, one of the best books in chemistry remains Antoine Lavoisier's *Elements of Chemistry* (1783, reprinted 1965). Although some of the terms might be unfamiliar, the writing and purpose remain clear to this day.

For information about specific scientists, the standard reference work is Charles Coulston Gillispie's *Dictionary of Scientific Biography* (1981). This massive, multivolume work was the collective work of almost everyone working in the field of history of science at the time. It remains the definitive source for biographical details about scientists from ancient times to the mid-twentieth century. A nice single-volume reference is *Collins Biographical Dictionary of Scientists* (1994), edited by Trevor I. Williams.

The historical study of alchemy has also increased dramatically in the past few years, driven in part by new research on Newton's interest in alchemy. For many years, E. J. Holmyard's *Alchemy* (1957, reprinted 1990) was the most commonly available introduction to the topic, but newer material such as

Bruce T. Moran's *Distilling Knowledge: Alchemy, Chemistry, and the Scientific Revolution* (2005) has greatly expanded the place of alchemy in the history of science. Stanton J. Linden's *The Alchemy Reader: From Hermes Trismegistus to Isaac Newton* (2003) offers the reader the original material with some helpful introductions.

In addition to traditional bibliographic sources, there are a number of interesting internet sources. Although Web sites change in the blink of an eye, the following sites are by organizations or individuals with long-term commitments to communicating material to the public. All the major chemical organizations have Web sites. News, historical material, and information about the profession of chemistry can be found on the following sites: American Chemical Society (www.chemistry.org), the Royal Chemical Society (www.rsc.org), and the International Union of Pure and Applied Chemistry (IUPAC) (www.iupac.org).

For historical and education material, there are two great sites at the Chemical Education Foundation (www.chemed.org) and the Chemical Heritage Foundation. (www.chemheritage.org). A large collection of alchemical material is maintained by Adam McClean at The Alchemy Web Site (www.levity.com/alchemy).

One of the most comprehensive and historically significant sites on the Web is the Nobel Prize site (nobelprize.org). This site has biographies of all the prize winners, their acceptance speeches, and historical and educational material on the topics that won prizes, as well as the history of the Nobel family and the prizes themselves. The site even lists the prize money awarded since the first prizes were given out.

Two sites about the periodic table are fun and useful. The first is the interactive periodic table of the elements maintained by the Los Alamos National Laboratory's Chemical Division (periodic.lanl.gov). Each element is described, along with physical constants and a brief history of its discovery. Although a few of the historical comments are not complete, the other information is very current. A site with a more historical orientation is Peter van der Krogt's Elementymology & Elements Multidict (www.vanderkrogt.net/elements/index.html). The site has an interactive periodic table, but van der Krogt is an historian with a strong interest in etymology, or the study of the origin of names. Each element's name is explained, along with alternatives that were used or proposed in the past. There are also sections that give the element names in 72 different languages.

BIBLIOGRAPHY

GENERAL HISTORIES OF SCIENCE

Bernal, J. D. *Science in History.* Harmondsworth: Penguin, 1969.

Bryson, Bill. *A Short History of Nearly Everything.* Toronto: Doubleday, 2003.

Ede, Andrew, and Lesley Cormack. *A History of Science in Society: From Philosophy to Utility.* Peterborough, Ont.: Broadview Press, 2004.

Heilbron, J. L., ed. *The Oxford Companion to the History of Modern Science.* Oxford: Oxford University Press, 2003.

Mason, Stephen F. *A History of the Sciences.* New York: Collier Books, 1962.

GENERAL HISTORIES OF CHEMISTRY

Brock, William H. *The Norton History of Chemistry.* New York: Norton, 1992.

Ihde, Aaron J. *The Development of Modern Chemistry.* New York: Harper & Row, 1964.

Jaffe, Bernard. *Crucibles: The Story of Chemistry from Ancient Alchemy to Nuclear Fission.* New York: Dover, 1976.

Levere, Trevor H. *Transforming Matter. A History of Chemistry from Alchemy to the Buckyball.* Baltimore: Johns Hopkins University Press, 2001.

Partington, J. R. *A History of Chemistry.* 4 vol. London: Macmillan, 1961–1970.

———. *A Short History of Chemistry.* New York: Dover, 1989.

THE ELEMENTS AND THE PERIODIC TABLE

Ball, Philip. *The Elements: A Very Short Introduction.* Oxford: Oxford University Press, 2004.

Clericuzio, Antio. *Elements, Principles, and Corpuscles: A Study of Atomism and Chemistry in the Seventeenth Century.* Dordreckt: Kluwer Academic, 2000.

Emsley, John. *Nature's Building Blocks: An A-Z Guide to the Elements.* Oxford: Oxford University Press, 2003.

Knight, David. M. *Atoms and Elements. A Study of Theories of Matter in England in the Nineteenth Century.* London: Hutchinson, 1967.

Morris, Richard. *The Last Sorcerers: The Path from Alchemy to the Periodic Table.* Washington, D.C.: Josephy Henry Press, 2003.

Nye, Mary Jo. *From Chemical Philosophy to Theoretical Chemistry: Dynamics of Matter and Dynamics of isciplines, 1800–1950.* Berkeley: University of California Press, 1993.

Strathern, Paul. *Mendeleyev's Dream. The Quest for the Elements.* New York: Berkley Books, 2000.

Stwertka, Albert. *A Guide to the Elements.* New York: Oxford University Press, 2002.

Weeks, Mary Elvira, with Henry M. Leicester. *Discovery of the Elements.* New York: Journal of Chemical Education, 1968.

PRIMARY SOURCES

Lavoisier, Antoine. *Elements of Chemistry.* Translated by Robert Kerr. New York: Dover, 1965. Also available electronically at galenet.galegroup.com at "The 18th-Century Collection Online."

Leicester, Henry M., and Herbert S. Klickstein. *A Source Book in Chemistry, 1400–1900.* Cambridge, Mass.: Harvard University Press, 1968

SCIENTISTS

Gillispie, Charles Coulston, ed. *Dictionary of Scientific Biography.* New York: Scribner, 1981.

Williams, Trevor I., ed. *Collins Biographical Dictionary of Scientists.* Glasgow: HarperCollins, 1994.

ALCHEMY

Holmyard, E. J. *Alchemy.* New York: Dover, 1957, repr. 1990.

Linden, Stanton J. *The Alchemy Reader: From Hermes Trismegistus to Isaac Newton.* New York: Cambridge University Press, 2003.

Moran, Bruce T. *Distilling Knowledge: Alchemy, Chemistry, and the Scientific Revolution.* Cambridge, Mass.: Harvard University Press, 2005.

INTERNET SOURCES

Adam McClean. The Alchemy Web Site. At www.levity.com/alchemy.

American Chemical Society. At www.chemistry.org.

Chemical Education Foundation. At www.chemed.org.

Chemical Heritage Foundation. At www.chemheritage.org.

International Union of Pure and Applied Chemistry (IUPAC). At www.iupac.org.

Nobel Prizes. At nobelprize.se.

Periodic Table of the Elements. Los Alamos National Laboratory's Chemical Division. At periodic.lanl.gov.

Peter van der Krogt. Elementymology & Elements Multidict. At www.vanderkrogt.net/elements/index.html.

Royal Society of Chemistry. At www.rsc.org.

INDEX

Abelson, Philip, 105, 148
Abu Bakr Muhammad ibn Zakariyya.
  See Al-Razi
Académie Royale des Sciences,
  57–58, 82
Academy of Plato, 13
Accelerator, linear, 100, 109
Actinium, 106, 144–45
Agricola, Georgius, 125, 142
Air pump, 49
Akap'ev, G. N., 153
Albertus Magnus, 35, 119
Alchemical laboratory, 40
Alchemists, 16, 19, 40, 43; Chinese,
  20, 22–23
Alchemy, 28, 43; and astrology, 21,
  36; and charlatans, 42; end of, 43;
  modern, 100
Alchymia, 41
Alcuin, 32
Al-Harrani. *See* Jabir ibn Hayyan
Al-kimiya (alchemy), 23
Alonso, J. R., 153
Alpha particles, 98
Al-Razi, 25–27, 28; list of substances, 26
Al-Sufi. *See* Jabir ibn Hayyan
Aluminum, 135
Americium, 105, 149
Analytical chemistry, 64
Anaximander, 10

Anaximenes, 10
Andreyev, A. N., 155
Antimony, 119
Arabic origins of English terms, 33
Arfvedson, Johan August, 133
Argon, 84, 106, 142
Aristotle, 14–19, 33, 45, 49
Armbruster, Peter, 154
Arrhenius, Carl Axel, 127
Arsenic, 119
Arthasastra, 27
Astatine, 148
Astatium, 148
Aston, Francis William, 97
Atomic number, 90
Atomic volume, 82
Atomic weight, 74
Atomism, 11, 17; Daltonian, 68–69,
  71, 90, 95
Atum, 2
Avicenna, 38
Avogadro, Amedeo, 78–79;
  hypothesis, 78; number, 94

Bacon, Roger, 36–37
Bacon, Sir Francis, 39
Baconianism. *See* Scientific method
Balard, Antoine-Jérôme, 135
Barium, 132
Bartholomew the Englishman, 34–35

Basilica Chymia, 41
Becher, Johann, 52
Becquerel, Henri, 93, 126
Berg, Otto, 147
Bergman, Torbern, 125
Berkelium, 150
Berthollet, Claude Louis, 59, 63, 70
Berzelius, Jöns Jakob, 64, 70, 74, 126,
　127, 128, 130, 134, 136, 137
Bismuth, 121–22
Black, Joseph, 54, 122
Blomstrand, Christian, 129
Bogomolov, Sergey L., 156
Bohr, Neils, 96, 103, 146, 153
Bohrium, 153
Bonds, 96
Book of Secrets, 35
Book of the Composition
　of Alchemy, 33
Book of the Secret of Secrets,
　The, 26
Boron, 132–33
Boyle, Robert, 48–50, 52, 57, 120
Bragg, Lawrence, 90
Braham, 2
Brain, C. K., 3
Brand, Hennig, 120
Brandes, J.F.W., 134
Brandt, Georg, 120
Brass, 7
Brauner, Bohuslav, 84, 149
Breeder reactors, 106
Brevium, 146
British Museum, 129
Bromine, 135
Bronze, discovery, 6
Bronze Age, 1
Brown, Robert, 94
Browne, C. I., 151
Brownian motion, 94
Buddhism, 21
Buklanov, German Vladimirovich, 156
Bunsen, Robert Wilhelm, 74–76,
　137, 138; burner, 75

Cadmium, 133–34
Caesium, 137
Calcium, 132
Californium, 150

Caloric, 61
Calorie (food), 61
Calorimeter, 61
Calx (oxidation), 53, 132
Cannizzaro, Stanislao, 78
Carbon, 115
Carbon-14 dating, 98
Carolingian Renaissance, 32
Cassiopeium, 146
Cathode rays, 92
Cavendish, Henry, 54, 55, 84, 124
Celestial realm, 13, 15
Celtium, 146
Ceramics, first use, 4
Ceria, 88; ceria series, 89
Cerium, 129–30
CERN. *See* European Centre for
　Nuclear Research
Cesium, 75, 137
Chadwick, James, 97
Chain reaction, 101
Chaptal, Jean Antoine, 122
Charlemagne, 32
Charleton, Walter, 48
Chelnokov, L. P., 153
Chelnokov, M. L., 156
Chemistry, origin, 45
Chepigin, Victor Ivanovich, 15
Chlorine, 123
Choppin, Gregory R., 151
Chromium, 64, 127–28
Clement, Nicolas, 133
Cleve, Per Theodor, 139
Cobalt, 120
Colbert, Jean-Baptist, 58
Cold War, 100, 107, 110
Collection of the Most Important
　Military Techniques, 23
Columbium, 129
Columbus, Christopher, 129
Combustion, theory, 49
Confuciansim, 21
Conservation of mass, 62
Constant change, concept of, 12
Copper, 116; mining, 5; smelting, 5
Coronium, 85
Corpuscle, 42,49, 92
Correspondence of Marianos the Monk
　with the Prince Khalid ibn Yazid, 24

Corson, Dale R., 148
Coryell, Charles D., 150
Cosmotron, 100
Coster, Dirk, 146–47
Cours de Chimie, 42
Courtois, Bernard, 133
Crawford, Adair, 126
Creation stories, 2–5; Chinese, 2;
    Egyptian, 2; Greek, 4; Hindu, 2;
    Hopi, 2; Japanese, 4–5
Croll, Oswald, 41
Cronstedt, Axel Fredrik, 121
Crookes, William, 92, 138; tube
    (electron discharge), 92
Cruikshank, William, 126
Curie, Marie Sklodowska, 93, 144,
    145, 149
Curie, Pierre, 93, 144, 145, 149
Curium, 105, 149
Cyclonium, 102
Cyclotron, 100, 109, 147

Dalton, John, 67–70, 71
Darmstadtium, 108, 154–55
Davy, Humphry, 70, 122, 123, 126,
    131, 132, 133, 134, 135, 142
Davy Medal, 86
Debierne, André-Louis, 144, 145
De caelo, 14
de Chancourtois, A. E. Béguyer, 77
Del Rio, Andrés Manuel, 128
Demarçay, Eugène-Anatole, 145
de Marignac, Jean Charles Galissard,
    140, 129
Demin, A. G., 153
Democritus of Abdera, 11, 17
de Morveau, Bernard Guyton, 135
Dephlogisticated air, 55, 84
De rerum natura, 17
Desaga, Peter, 75
Descartes, Rene, 46–47, 50
Désormes, Charles-Bernard, 133
de Ulloa, Antonio, 121
de Zubice, Faust de Elhuyar y, 125
de Zubice, Juan José de Elhuyar y, 125
Diamond, H., 151
Die modernen Theorien der Chemie,
    78, 82
Dilithium, 111

DNA, 93, 120
Döbereiner, Johan, 76–77
Doctrine of Phlogiston, The, 59
Donets, E. D., 107
Dorn, Friedrich Ernst, 145
Druin, V. A., 152, 153
Dubnium, 152–53
Dulong, Pierre Louis, 74
"Dynamid," 95

Einstein, Albert, 95, 150
Einsteinium, 106, 150–51
Eka-boron, 83
Eka-silicon, 83
Ekeberg, Anders Gustaf, 88, 129
Electrical discharge tube.
    See Crookes, William
Electron, 90, 92, 96, 97
Electron death spiral, 95
Elixir of life, 45
Epicureans, 17
Epicuro-Gassendo-Charletoniana, 48
Epicurus of Samos, 17, 47
Erasmus, 38
Erbium, 136–37
Eskola, K., 153
Eskola, P., 153
Essai de statique chimique, 70
Ether, 13, 85
European Centre for Nuclear Research
    (CERN), 100
Europium, 145
Euxenium, 146
*Experiments and Observations on
    Different Kinds of Air*, 55

Fajans, Kasimir, 146
Fefilov, B. V., 153
Fermium, 106, 151
Fields, P. R., 150, 151
Fire, first use, 3, 4
Fission, 101
Fixed air, 54
Fixed proportions, law of, 64
Flerov, Georgy Nikolaevich, 107,
    152, 153
Flerov Laboratory of Nuclear Reactions,
    Russia, 108, 155, 156
Fluorine, 142

Folger, H., 154
Four causes, 15, 16
Fourcroy, Antoine, 59
Four element system (earth, water, air, fire), 13
Four qualities, 15
Francium, 102, 147
Frankland, Edward, 143
Franklin, Benjamin, 54
French Revolution, 63
Fried, Sherman M., 151
Frisch, Otto, 101
Fusion, 106

Gadolin, John, 88, 127
Gadolinium, 140
Gahn, Gottlieb Johan, 123
Galen, 32, 38
Gallium, 82, 138–39
Gas, 61
Gassendi, Pierre, 47–48
Gavrilov, K. A., 152
Gay-Lussac, Joseph Louis, 71, 78, 133
Geb, 2
Geber. *See* Jabir ibn Hayyan
Geiger, Hans, 95, 98
Geissler, Heinrich, 91
Geist: pinque, 42; sylvestre, 42
Genesis, 3, 5, 7
Geometric solids, 10, 11
Germanium, 83, 141
Gesellschaft für Schwerionenforschung, Germany, 154, 155
Ghiorso, Albert, 149, 150, 151, 152, 153
Giesel, Fritz, 145
Gikal, Boris N., 156
Gilbert, Ludwig Wilhelm, 131, 135
Glass, discovery, 6
Glendenin, Lawrence E., 102, 150
Gluon, 100
Göhring, Otto H., 146
Gold, 116
Goldstein, Eugen, 92
Gorshkov, Vladimir A., 156
Great Britain Atomic Energy Authority, 146
Gregor, William, 127
Gulbekian, Georgy G., 156

Gunpowder, 23
Gunpowder and Saltpeter Administration, 58
Guyton de Morveau, Louis-Bernard, 59

Hafnium, 106, 146
Hahn, Otto, 100, 146, 153
Hahnium, 108, 153
Hales, Stephan, 53
Half-life, 98
Haniyasu-hime, 5
Hanoyasu-hiko, 5
Haroon al-Rashid, 25
Harris, J., 153
Harris, John W. K., 1
Harvey, Bernard G., 151
Hassium, 154
Hatchett, Charles, 129
Helium, 84, 142–43
Henry, William, 68
Heraclitus of Ephesus, 12
Herakles (Hercules), 4
Hermann, Carl Samuel, 134
Hertz, Heinrich, 92
Heßberger, Fritz Peter, 154, 155
Higgins, Gary H., 150, 151
High Frequency Spectra of the Elements, The, 91
Hingmann, R., 154
Hirsch, A., 151
Hisinger, Wilhelm, 64, 130
Hittites, 7
Hjelm, Jacob, 124
Hofmann, Sigurd, 154, 155
Holmberg, O., 139
Holmium, 139
Hooke, Robert, 48, 52
Hope, Thomas Charles, 64, 126
Huizenga, John R., 151
Hulet, E. Kenneth, 153
Hunter, Matthew A., 127
Hydrogen, 124; structure, 97

Ideal Forms, 13, 14
Impossibility of change, 12
Indium, 75, 138
Indonesia, 20
International Union of Pure and Applied Chemistry (IUPAC),

108, 109, 112, 129, 135, 146,
    152, 153, 154, 155, 156
International Union of Pure and
    Applied Physics (IUPAP), 109
Iodine, 133
Iridium, 130–31
Iron, 7, 118; smelting, 7, 21
Iron Age, 1
Ising, Gustaf, 100
Islam, 20, 24
Isotope, 97
Itkis, Michael Grigorievich, 156
IUPAC. *See* International Union
    of Pure and Applied Chemistry
IUPAP. *See* International Union
    of Pure and Applied Physics

Jabir ibn Hayyan, 24–25, 33;
    four natures theory of, 25
James, Ralph A., 149
Janik, Rudolf, 155
Janssen, Pierre-Jules-César, 142
Jargonium, 146
Jing, Emperor, 22
Joint Institute for Nuclear Research,
    152, 153
Joliot, Jean-Frédéric, 99
Joliot-Curie, Irene, 99
Journal of the Chemical Society, 77
*Journal of the Russian Chemical
    Society,* 81
Junine, Claude Geoffroy, 122

Kabachenko, Anatoly Petrovich, 156
Kagutsuchi-no-Kami, 4
Kalium, 131
Kanayama-biko, 4
Kanayama-hime, 4
Karlsruhe Congress, 80
Kautilya, 27
Kekulé, Friedrich August, 71–72
Keller, J., 154
Kennedy, Joseph W., 148
Khalid ibn Yazid, 24, 33
Kharitonov, Yu. P., 153
Kirchhoff, Gustav Robert, 74–75,
    137, 138
Kirwan, Richard, 126
Kitaibel, Paul, 125

Klaproth, Martin Heinrich, 64–65, 67,
    88, 125, 126, 128, 130–32, 134
Klaprothium, 134
Klaus, Karl Karlovich, 89, 137
Kolesov, I. V., 153
Kroll, William, 127
Krypton, 143
Kurchatovium, 107
Kuznetsov, V. I., 152

Lamy, Claude-Auguste, 138
Langmuir, Irving, 96, 97
Lanthanum, 90, 106, 136
Lao-tsu, 20
Laplace, Pierre Simon, 61
Larsh, Almon E., 152
Latimer, R. M., 152
Lavoisier, Antoine Laurent, 57–63,
    65, 67, 70, 74, 88, 120, 123,
    124, 135
Lavrentiev, Anton Yurievich, 156
Law of Octaves, 77
Lawrence, Ernest O., 100, 152
Lawrencium, 107, 152
Lead, 7, 117
*Leçons élémentaires d'histoire naturelle
    et de chimie,* 59
Lecoq de Boisbaudran, Paul (François)
    Émile , 82–83, 138, 140
Leino, Matti, 154, 155
Lémery, Nicolas, 42, 119
Leucippus of Miletus, 11
Lewis, Gilbert Newton, 96, 97
Libavius, Andreas, 41
Light, wave/particle problem, 51–52,
    92, 93
Lithium, 76, 133
Lobanov, Y. V., 152, 153
Lockyer, Norman, 143
Lodestone, 118
Lougheed, R. W., 153
Löwig, Carl, 135
Lucretius, 17
Lutecia, 146
Lutetium, 106, 145–46
Lyceum of Aristotle, 14

Mackenzie, Kenneth R., 148
Magic, 5, 35

Magnesium, 122
Malyshev, Oleg Nikolaevich, 156
Manganese, 123
Manhattan Project, 101–2, 105–6,
     142, 149
Manning, W. M., 151
Marggraf, Andreas Sigismund, 120
Marinsky, Jacob A., 102, 150
Maryanos (Marienus), 24
Mass, concept of, 42
Mass spectroscopy, 97
Masurium, 147
Matter theory, 1, 2; Chinese, 21;
     Greek, 10
McMillan, Edwin M., 105, 148
Mech, J. F., 151
Meissner, W., 134
Meitner, Lise, 100, 141, 146
Meitnerium, 154
Mendeleev, Dimitri Ivanovich, 77,
     79–86, 89–90, 138, 139, 141, 146,
     147, 148
Mendelevium, 106, 151
Mercury, 118
Meteorologica, 33
Méthode de nomeclature chimique, 60
Meyer, Julius Lothar, 78–79, 82
Minerals creation, ancient theory, 6
Moissan, Ferdinand-Frédéric-
     Henri, 142
Molybdenum, 124
Morgan, Leon O., 149
Mosander, Carl Gustav, 127, 136–37
Mosandrum, 140
Moseley, Henry, 90
Motion and the elements, 15
Münzenberg, Gottfried, 154, 155
Muon, 100

Naming elements, modern system, 108
Natural philosophy, 41; Greek
     origins, 9
Nature, 101
Neodynium, 141
Neon, 143
Nephthys, 2
Neptunium, 105, 148
Neutron, discovery, 97
New Alchemy, The, 98

New Atlantis, The, 39
*New Experiments Physico-
     Mechanicall Touching on the
     Spring of Air*, 48
Newlands, John, 77, 86
New System of Chemical Philosophy, 68
Newton, Sir Isaac, 42, 48, 50–51,
     71, 73, 92
Nickel, 121
Nicomachus, 14
Nielsbohrium, 108
Nigrium, 146
Nilson, Lars Frederick, 83, 139
Nimov, Victor, 154
Niobium, 129
Nitrogen, 84
Nitschke, J. M., 153
Nobel, Alfred, 151
Nobelium, 107, 151–52
Nobel Prize, 90, 107; chemistry, 94;
     physics, 93
Noble gases, 85
Noddack, Walter, 147
Nomenclature, 60
Norium, 146
Norwegium, 146
Novum Organum, 39
Nuclear bomb, 98, 106, 153; making
     of, 102
Nuclear reactor, 106
Nucleus, 95
Nurmia, M., 153
Nut, 2

Oecolampadius, 38
Oganessian, Yuri Tsolakovich,
     152, 156
Ogawa, Masataka, 147
On the Properties of Things, 34–35
Opticks, 51
Ørsted, Hans Christian, 135
Osann, Gottfried Wilhelm, 137
Osiris, 2
Osmium, 130
Ostranium, 146
Owens, Robert B., 145
Oxygen, 59, 84, 122–23

Palladium, 89, 130

Pan Gu, 2, 3
Paracelsus, 37–39, 41, 49, 118, 120
Parmenides, 12
Pauling, Linus, 94
Péligot, Eugène, 126
Penglai, goddess of, 22
Perelygin, V. P., 152
Perey, Marguerite, 102, 147
Periodic Law, 81
Periodic table, 81, 97, 113
Perrin, Jean, 94–95
Petit, Alexis Thérèse, 74
Philosopher's Stone, 16, 42, 45, 100
Philosophiae Epicurus Syntagma, 48
*Philosophiae naturalis principia
    mathematica*, 50
Philosophical Magazine, 91
*Philosophical Transactions of the
    Royal Society*, 50
Phlogisticated air, 84
Phlogiston, 53, 59
Phophorus, 120
Physicae subterraneae, 52
Physics, 33
Pitchblende, 64, 93
Planck, Max, 96
Platinum, 88, 89, 121
Plato, 13–14, 19, 33–34, 39
Platonic solids, 13
Plotko, V. M., 152
Plücker, Julius, 92
Plum pudding model, 95
Plutonium, 105, 106, 148–49
Pneumatic chemistry, 52
Pneumatic trough, 53–54
Polonium, 93, 144
Popeko, Andre Georgievich, 155, 156
Poppensieker, Klaus, 154, 155
Potassium, 76, 131
Potassium-40 dating, 98
Praja-pati, 2
Praseodymium, 141
Price, James, 43
Priestley, Joseph, 54–56, 59, 123
Prime matter, 2
Primordial germs, 17
Principia, 50
Principles of Chemistry, 80
Prometheus, 4

Promethium, 102, 149–50
Protactinium, 146
Proton, 90, 97; discovery, 96
Proton bombardment, 98
Proust, Joseph Louis, 64
Ptolemy, 73
Pyle, G. L., 151
Pythagoras, 10
Pythagoreans, 10, 11

Quanta, 96
Quantum physics, 96, 99
Quill, L. L., 102
Quintessence, 13

Radioactivity, 98; discovery of, 93
Radium, 93, 144
Radon, 145
Raisin muffin model, 95
Ramsay, William, 84–85, 142, 143, 144
Rare-earth elements, 64, 83–84,
    88, 106
Rasarnava (Treatise on Metallic
    Preparations), 28
Rayleigh, Lord, 84–85, 142
Reflections on Phlogiston, 58
Reich, Ferdinand, 138
Reisdorf, Willibrord, 154
"Relation between the Properties
    and Atomic Weights of the
    Elements," 81
Republic, 39
Rhenium, 147
Rhodium, 89, 131
Richter, Harold G., 150
Richter, Hieronymus Theodor, 138
Robert of Chester, 33
Roentgenium, 109, 155
Rohach, Jozef, 156
Roloff, J.C.H., 134
Romans, 7
Röntgen, Anna Bertha, 92
Röntgen, Wilhelm Conrad, 92, 155
Rose, Valetin, 88, 129
Royal Society, 43, 57–58, 86;
    founding, 50
Rubidium, 75, 137–38
Ruthenium, 137
Rutherford, Daniel, 122

Rutherford, Ernest, 90, 95, 97, 98, 102, 145
Rutherfordium, 108, 152, 153

Sagaidak, Roman Nikolaevich, 156
Sahm, Christoph-Clemens, 154
Salt, 132
Samarium, 140
Samaskaras, 28
Saro, S., 155
Saturnian atom, 95
Scaliger, Julius Caesar, 121
Scandium, 83, 139
Sceptical Chymist, The, 49
Scharff-Goldhaber, Gertrude, 106, 109
Scheele, Carl Wilhelm, 123–25, 132
Schmidt, Karl-Heinz, 154
Schneider, J.R.H., 154
Schneider, W.F.W., 154
Schött, H.-J., 154
Schrödinger, Erwin, 99; cat and diabolical device, 99
Schroeder, Johann, 119
Scientific method, 39–40
Seaborg, Glenn T., 105, 106, 107, 108, 148, 149, 150, 151, 153
Seaborgium, 108, 153
Sefström, Nils Gabriel, 128
Segrè, Emilio G., 147, 148
Selenium, 134
Semaw, Sileshi, 1
Seolfor, 116
Seth, 2
SHE. See Superheavy elements
Shu, 2
Siedentopf, Heinrich F., 94
Sikkeland, Torbjørn, 151, 152
Silicium, 134–35
Silicon, 134
Sillent, A., 3
Silver, 116
Skobelev, Nikolai Konstantinovich, 153
Smith, H. Louise, 151
Smith, J. Lawrence, 140
Sodium, 76, 131–32
Soret, Jacques-Louis, 139
Sotuknang, 2
Specific gravity, 82

Spectroscope, 75, 84–85
Spence, R. W., 151
Speusippus, 14
Stahl, Georg, 53
Stars, 75, 106
Star Trek, 111
States of matter, 10
Strassmann, Fritz, 100
Street, Kenneth, Jr., 150
Stromeyer, Friedrich, 134
Strontia, 64, 126
Studier, Martin H., 150, 151
Subatomic particle, 100
Sulfur, 117
Superheavy elements (SHE), 106, 109
Sylva Sylvarum or a Naturall Historie in ten Centuries, 39
Sylvius, Franciscus, 41
Synchrotron, 100
Système des connaissances chimiques, 59
Szilard, Leo, 101

Tacke, Ida Eva, 147
Taiowa, 2
Tantalum, 90, 129
Tantrism, 28
Taoism, 20, 24
Tao te Ching (Book of the Way), 20
Tarkasamgrahadeepika, 27
Technetium, 147–48
Tefnut, 2
Telluric helix, 77
Tellurium, 64, 88, 124
Teniers the Younger, David, 40
Tennant, Smithson, 89, 130
Terbium, 136
Terra fluida (mercurious earth), 52
Terra lapidea (vitreous earth), 52
Terra pinguis (fatty earth), 52
Terrestrial realm, 13, 15
Thales of Miletus, 10, 112
Thallium, 75, 138
Thermonuclear weapon, 150
Tholden, Johann, 119
Thompson, Stanley G., 150, 151
Thomson, J. J., 92
Thorium, 93, 136
Thorium-232 dating, 99

Three earths theory, 52
Three principles system of
    Paracelsus, 38
Thulium, 139
Thuma, B., 154
Timaeus, 13, 33–34
Tin, 7, 117
Titanium, 64, 88, 106, 127
Traité élémentaire de chimie (Elements
    of Chemistry), 60, 62
Transmutation, 31, 39–41, 43;
    radioactive, 98, 99
Transubstantiation and transmutation,
    31, 34
Travers, Morris W., 85, 143, 143, 144
Tretiakova, S. P., 152
Triad system, 76
Tubal-Cain, 7
Tungsten, 125

Ultramicroscope, 94
Ununquadium, 156
Unununium, 155
Urbain, Georges, 146
Uranium, 64, 88, 93, 125
Uranium-235, 97, 102
Uranium-238, 102

*Valence and the Structure of Atoms and
    Molecules*, 96
Vanadium, 128
Van de Graaf, Robert, 100; generator 100
Van Helmont, Jan Baptist, 41–42, 46, 61
Vauquelin, Nicolas-Louis, 128
Vedic writing, 27
Vermeulen, Detlef, 154
Vitalism, 27
Void, 14
Volta, Alessandro, 70; Voltaic pile, 70
Volumetric analysis, 71
von Hevesy, György Karl, 146

von Hohenheim, Philippus Aureolus
    Theophrastus Bombast. *See*
    Paracelsus
von Reichenstein, Joseph Müller
    Freiherr, 125
von Welsbach, Carl Auer, 140, 141, 146

Wahl, Arthur C., 148
Walton, J. R., 151
Watson, William, 121
Wei Po-Yang, 22–23
Wheeler, John, 106, 109
Willow tree experiment, 41
Winkler, Clemens Alexander,
    83, 141
Winthrop, John, 129
Wöhler, Friedrich, 128, 135
Wollaston, William Hyde, 89, 130, 131
World War I, 90, 96, 123
World War II, 102, 105
Woulfe, Peter, 125
Wu, Emperor, 22

Xenon, 106, 143–44
X-ray, 93; diffraction, 90; discovery, 92

Yeremin, Alexander Vladimirovich,
    155, 156
Yin yang, 21
Young, Thomas, 92
Ytterbium, 140
Ytterby, Sweden, 84, 88, 127, 136, 140
Yttria, 88; series, 89
Yttrium, 127

Zeng Gongliang, 23
Zeno, 12; paradox, 12
Zinc, 119–20
Zircon, 64
Zirconium, 88, 106, 126
Zsigmondy, Richard, 94

About the Author

ANDREW EDE is an Assistant Professor in the Department of History and Classics at the University of Alberta. His most recent book was *A History of Science in Society* with Lesley Cormack.